Toxic Matters

Under the Sign of Nature: Explorations in Ecocriticism
Serenella Iovino, Kate Rigby, and John Tallmadge, Editors
Michael P. Branch and SueEllen Campbell, Senior Advisory Editors

Toxic Matters

NARRATING ITALY'S DIOXIN

Monica Seger

UNIVERSITY OF VIRGINIA PRESS
CHARLOTTESVILLE AND LONDON

University of Virginia Press
© 2022 by the Rector and Visitors of the University of Virginia
All rights reserved
Printed in the United States of America on acid-free paper

First published 2022

ISBN 978-0-8139-4835-5 (hardcover)
ISBN 978-0-8139-4836-2 (paperback)
ISBN 978-0-8139-4837-9 (ebook)

9 8 7 6 5 4 3 2 1

Library of Congress Cataloging-in-Publication Data is available for this title.

Cover art: From the documentary film *Buongiorno Taranto,* Italy, 2014 (used by permission of the director, Paolo Pisanelli, Big Sur/OfficinaVisioni); *background images:* iStock.com/xxmmxx; iStock.com/Brett_Hondow

Contents

Acknowledgments   vii

Introduction: Narrating Italy's Dioxin   1

1. Seveso: Making Sense   21
   First Person: Members of the Circolo Legambiente Laura Conti Seveso   32
2. Seveso Stories, or The Importance of Laura Conti   37
   First Person: Massimiliano Fratter   62
3. Taranto: Past, Present, Future   63
   First Person: Daniela Spera   73
4. Toxic Tales: Mapping Plurivocality, Mostly on Page   77
   First Person: Angelo Cannata   101
5. On-screen Engagements: Mapping Plurivocality (and Not)   105
   First Person: Vincenzo Fornaro   137
6. Reading Landscapes: Back to the Land in Seveso and Taranto   141
   First Person: The Ammostro Artist Collective   174

Notes   179
Bibliography   197
Index   207

Acknowledgments

I believe strongly in the power of collective work, and this book proves no exception. The project behind it came about thanks only to many years of conversation with dear colleagues, whether held via conference panels, during meandering walks, through email exchanges, or, in its final year of gestation, over Zoom lunches. I am indebted to a dynamic and ever-expanding community of eco-Italianists who have provided me with both inspiration and encouragement. Damiano Benvegnù, Enrico Cesaretti, Laura Di Bianco, Matteo Gilebbi, Serenella Iovino, and Elena Past have been especially present in these conversations, and I cannot thank them enough. Laura and Elena's steady support was truly precious as the book neared production. I am also deeply grateful to KT Thompson for enriching conversations and friendship beyond the realm of Italian studies, to my amazing students and colleagues at William & Mary, and to all those friends and scholars whose paths I have crossed over the years, and whose work inspires my own.

Toxic Matters was deftly guided on its course by my editor at the University of Virginia Press, Angie Hogan, after having initially been welcomed there by Boyd Zenner. I am grateful to both of them, as well as to Ellen Satrom and her production team at UVaP, and the Under the Sign of Nature series editors, Serenella Iovino, Kate Rigby, and John Tallmadge. In addition, I thank Anna Frare and Renée Anne Poulin for their excellent transcription and translation of interviews, and Kate Blackmer for the fantastic map. I am also grateful to the press-appointed readers of my manuscript, who were both so generous in their engagement with my work and offered invaluable suggestions for improvement. I hope to have done them justice. I thank Allison Cooper, Danielle Hipkins, Catherine O'Rawe, and Dana Renga for important early feedback on the project, and Cristina Zagaria for a lovely afternoon interview in Naples, complete with pizzas and ocean views. A thanks as well to Rob Nixon, as I would not have found my way to Italy's dioxin narratives had he not suggested I look up Laura Conti many years ago.

Portions of this book have previously appeared in the journal *ISLE: Interdisciplinary Studies in Literature and Environment* 24, no. 1 (2017) and the edited volumes *Encounters with the Real in Contemporary Italian Literature and Cinema* (Cambridge Scholars Publishing, 2017) and *Italy and the Environmental Humanities: Landscapes, Natures, Ecologies* (University of Virginia Press, 2018). I thank the respective editors and publishers for kindly allowing me to reprint my work here. I also thank William & Mary for financial support toward both research and production, Trudy Hale at The Porches for providing such a special place for writers, and the organizers and coparticipants of the 2017 workshop on Radical Hope held at the Rachel Carson Center in Munich.

I am lucky to be endlessly supported by my loving family members, who have always expressed enthusiasm for my scholarship and never asked when a project was going to be complete. To them, too, I say thank you. I am also grateful for animal companions Luna and Zucchi, who supplement my life with quiet petting sessions and adventure walks, and especially for my partner, Clint, who cheers on my work while reminding me of the sweetness of life beyond it. My deepest gratitude goes to the activists, artists, and storytellers in Seveso and Taranto who have so generously shared time with me over the past many years: Angela Alioli, Gemma Beretta, Angelo Cannata, Vincenzo Fornaro, Max Fratter, Noel Gazzano, Alessandro Marescotti, Grace Zanotto, the artists of Ammostro, the members of Circolo Legambiente Laura Conti Seveso, and the tireless Daniela Spera.

Toxic Matters

Introduction
Narrating Italy's Dioxin

This is a book about the relationships between large-scale industry, daily domesticity, and narrative practice. As such, it is an investigation into the factory-proximate interplay between bodies, soil, and the wealth of dynamic matter that passes in between; a meditation on how all of the above is complicated by the strange workings of time, with presents dominated by past actions, and futures unclear; and, most of all, a study and celebration of the function of story. It considers what happens on levels both cognitive and emotional, concerning both understanding and affect, when someone crafts a story, when others retell it, and when all of us who comprise its audience experience it through reading, watching, and other receptive practices. This book proposes that narrative, in all of its many forms, allows storytellers and audiences alike to make sense of changing environmental and human health realities as they shape the worlds both around and within—especially when such changes are not immediately obvious or predictable.

The following chapters focus on contemporary Italy as their primary site of interest, and on the chemical compound known as dioxin(s) as the primary matter linking all of the soil, bodies, and stories discussed herein. Were this book about any number of other places in the world, inundated with any number of other toxins, I suspect its fundamental argument would largely be the same: narrative engagement facilitates deeper understanding of environmental and human (or animal) health crisis—what I will generally refer to as "eco-corporeal crisis" in the pages to come. The path to get there, however, would take a different shape, its surface covered with different dirt molded by a different set of roots. That is to say, the primary themes of toxicity and narrative at stake in this book concern all of us throughout the lived world, whereas the coordinates that shape them are often particular to place.

And the places of interest here are exactly two. The first, the northern Italian town of Seveso, is located fifteen kilometers northeast of Milan. It was home in 1976 to one of the greatest-known releases of dioxin into the atmosphere worldwide, from the ICMESA chemical factory. First opened in 1945, ICMESA (Industrie Chimiche Meda Società Azionaria), was a small subsidiary of the Swiss pharmaceutical and fragrance manufacturer Givaudan, which was itself incorporated by the multinational pharmaceutical company La Roche in the early 1960s. On July 10, 1976, one of ICMESA's chemical reactors overheated and ruptured, emitting a large white cloud of dioxin-heavy substances that lingered over Seveso for twenty minutes or more before dispersing into the atmosphere. As I detail in chapter 1, this event was followed by an extended period of official silence and, eventually, the dissemination of conflicting information regarding local health risks. During this time, hundreds of human residents fell ill as a result of exposure, and thousands of animals died. The years after the disaster were no easier, as the community struggled with the issue of elective abortion in the wake of exposure; questioned various acts of remediation both financial and ecological; and experienced lingering uncertainties regarding corporeal and environmental health.

The second is the southern coastal town of Taranto, home to the Ilva steelworks since 1964. Ilva-Taranto is the largest steelworks in all of Europe, with a nominal production capacity of eight million tons per year. As discussed in chapter 3, the steelworks has been under intense scrutiny for years, due to a long-running failure to comply with European environmental standards for industrial emissions. In 2013, the European Commission began infringement proceedings against Italy for its failure to ensure that Ilva adhere to European Union (EU) legislation. Later that year, Ilva managers were indicted and the steelworks placed under special administration of the Italian government. On November 1, 2018, ownership of the steelworks was transferred to multinational ArcelorMittal, the world's largest steel manufacturer. Since that time, the steelworks has been referred to as both "Ex Ilva" and "Ilva" in the press; I use the latter here for simplicity's sake. Emissions from the Taranto plant include harmful minerals, metals, and an extraordinarily high level of dioxin (a 2005 study reported that Ilva emitted 8.8 percent of the dioxin produced in all of Europe).[1]

There are obvious ways in which the cases present two very different scenarios, especially when it comes to the question of time. The residents of Seveso experienced a chemical explosion, a case of intense immediate

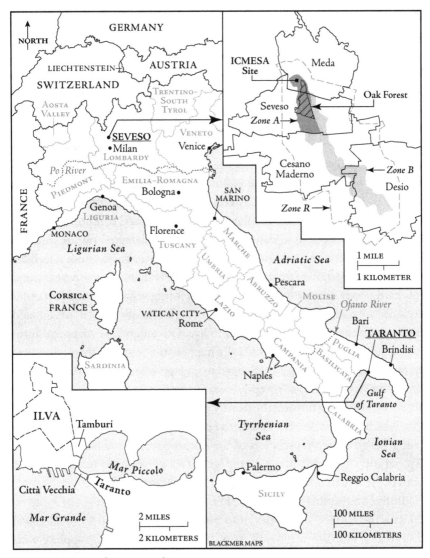

Italy, with insets of Seveso and Taranto

impact followed by an extended period of recovery and complicated by delayed notice from public officials, as well as often sensationalist coverage in the national media. The residents of Taranto, on the other hand, have been subject to a very long-term quotidian release of chemicals into the air. They have felt the "slow violence" that Rob Nixon theorized in his 2011 book of the same name, the sort of gradual destruction to the more-than-human environment and the bodies in its midst that rarely makes its way onto the nightly news.[2] What's more, Seveso has undergone a process of environmental reclamation and a certain measure of community-based reckoning in recent decades, especially at the former ICMESA factory site itself. Conversely, Taranto's massive steelworks is still actively producing both steel and toxic emissions, and public opinion differs widely about the factory's role in the local economy and the possibility of a future without Ilva.

Despite these temporally rooted differences, key coordinates place the crises of Seveso and Taranto in clear dialogue with one another. Most immediately, although other toxic pollutants were emitted by ICMESA, and still are by Ilva, dioxin is routinely considered to be the most destructive of those pollutants in both cases. Like many toxic chemical compounds, dioxin is nearly imperceptible in daily life. It cannot be detected by the unaided human eye, and it bears no noticeable smell or taste. Furthermore, as I discuss in greater detail below, it lingers in land and bodies for many years, and it is hard to say with certainty what effects it might have on any particular creature's health, or when. The residents affected by both ICMESA and Ilva's emissions know well what it is to live with an unclear corporeal futurity, unsure of what might eventually happen as a result of their exposure. At the same time, they are challenged to engage what Barbara Adam calls a timescape perspective, a "receptiveness to temporal interdependencies and absences," thinking beyond the overtly empirical and physically obvious "to that which is invisible and outside the capacity of our senses."[3] In this current geological age that scholars of the Anthropocene have labeled a "Great Acceleration," slowing down to engage such a perspective, to contemplate both the less than obvious and the long-term, is no small task.

Both communities have also had to confront the question of what to do with the physical territories most impacted by their respective chemical disasters. While Taranto is still, as of this writing, actively grappling with Ilva's present and future, residents in its proximity have already been forced to make difficult decisions about how to interact with a heavily polluted landscape, which has negatively impacted once-thriving fishing and agricultural industries. More than forty years after the ICMESA disaster and subsequent factory closure, the residents of Seveso and nearby communities have instead had ample time to respond—a long and complicated process. For many, this has included feelings of disgrace and anger, as well as a dismissal of potential danger from dioxin exposure. For others, response has taken the shape of a memorial park, including a community archive and nature preserve, built on the former ICMESA site. Though it has its complications, the remediative park project speaks directly to Lawrence Buell's 1998 hypothesis that people would increasingly come to "visualize humanity in relation to environment . . . as collectivities with no alternative but to cooperate in acknowledgement of their necessary, like-it-or-not interdependence."[4] The park recognizes the past uses of the land on which it rests, the ways in which that physical space hosted harm to humans (due, of course, to human action), and the fact that humans find joy and health by exerting themselves in open green spaces. The various responses displayed in Seveso, explored throughout the coming chapters, might serve as models, both positive and negative, for today's Taranto.

Finally, and most significantly for this book, both cases have inspired a range of creative narrative texts. Through novels and films, theatrical dramas and works of visual art, residents, witnesses, and interpreters have responded to Seveso and especially Taranto in a vast array of expressive modes—often seeking to chronicle the burgeoning and largely imperceptible health realities imposed by dioxin. Authors, filmmakers, and other artists have offered stories of place and community as a means to coping and comprehension, whether through sharing factual information about events underway or exploring anxieties about potential futures. They have also used story as a means of resistance, confirming Marco Armiero's claim that "to narrate means to counter-narrate."[5] Environmental injustice, slow eco-corporeal violence, is enacted not only through physical events like toxic emissions but also through official narratives that serve to "eradicate any other possible alternative, that impose an official truth."[6] By telling the many stories of Seveso and Taranto, individual and collective, in different voices and registers, using both realist and speculative modes, the narrators considered in this book together provide more nuanced and multiaxial views of events than those offered by state officials and mainstream media, just as they confront what Nixon would call the "representational challenges" posed by dioxin.[7] This is particularly the case regarding the ongoing crisis at Taranto, which has inspired a significantly wider array of creative texts than has Seveso. As such, I dedicate more attention to Taranto than to Seveso in the following chapters, while always recognizing the latter as a fundamental interlocutor.

I offer below a brief primer on dioxin, as it, along with narrative, serves as the primary connective tissue for my study. First, though, let me be clear that while it plays an essential role, dioxin is not the culprit for the various forms of harm wrought in Seveso and Taranto. Although I take an eco-materialist stance in recognizing dioxin's (narrative) agency, I do not mean to suggest that dioxin is the primary locus of toxicity in Seveso or Taranto; rather, it is a particularly dynamic means through which toxicity travels. Toxicity is a complicated constellation, one built into social structures at the level of law through exposure limits and chemical thresholds.[8] What's more, as Max Liboiron and colleagues write, toxicity is more than just "wayward particles behaving badly" and "harm at the cellular level." It is also "a way to describe a disruption of particular existing orders, collectives, materials and relations."[9] When we speak of toxicity, this state of disruption in existing orders and relations, we necessarily (although perhaps unknowingly) speak of the effects of a wide range of inequities "premised upon and reproduced by systems of colonialism, racism, capitalism, patriarchy, and other structures that require land and bodies as sacrifice zones."[10] The eco-corporeal toxicity that occurs in places like Seveso and Taranto, where low-income workers live closest to the factories that emit chemical by-products into air and water, and corporations have greater legal power than individual families, is a reflection of neoliberal society at large.

The harm suffered by land and bodies at Seveso and Taranto results from (mis)actions at and surrounding ICMESA and Ilva. It also results from the norms and expectations that did or indeed still do support those organizations, and that dictate who lives in closest proximity to factory grounds, how information is dispersed, why people feel they must choose between employment and health, and so on. Dioxin is not the culprit in Seveso and Taranto (to be clear: Hoffman LaRoche and Ilva are, as are the social structures that support the existence of those entities), but it is the common material agent through which illness and environmental imbalance did or still do travel. By following its physical paths and the ways in which it connects industry, environment, and living beings in both Seveso and Taranto, we are able to chart the existence of the sort of factory-proximate eco-corporeal crises that exist worldwide.

In its pure form, dioxin exists as minuscule solids or crystals lacking in color and odor. It usually spreads into the environment in combination with other substances such as ash, soil, or the leaves of plants, although some of the dioxin released from a chemical process may also be found in air or water, in a vapor or dissolved state, blending into the greater atmosphere. Dioxin can enter human and animal bodies when we breathe contaminated air or when we contact contaminated soil. Most often, however, it enters our bodies by moving through the food chain, increasing at each step in a process known as biomagnification.

Dioxin comes into being as a by-product of making other chemicals, manufacturing pesticides, burning forests and trash, or bleaching pulp and paper. It is often linked to furans and dioxin-like polychlorinated biphenyls (PCBs), due to their similar toxicity and chemical characteristics. Chlorinated dibenzo-p-dioxins (CDDs) actually comprise many chemically related compounds. The Agency for Toxic Substances and Disease Registry (ATSDR) notes that there are seventy-five different compounds commonly referred to as polychlorinated dioxins, and that each compound's particular name reflects the number and position of the chlorine atoms that it contains. The dioxin known as 2,3,7,8-tetrachlorodibenzo-p-dioxin, or 2,3,7,8-TCDD, is one of the most toxic dioxins, as well as the one that has received the most attention from science and mainstream media. It is also the dioxin that was released at the ICMESA plant in Seveso in 1976 and that has been found in abundance in the breast milk of Tarantine women.[11] For this reason, and in keeping with the tendency of both mainstream media and scientific fact sheets written for the general public, I use the singular "dioxin" here.

At a 1992 U.S. government hearing, then congressional representative Donald M. Payne of New Jersey declared, "Dioxin is a word that is well known but little understood."[12] I have found in my research that this still holds true to a certain extent. For one thing, it is not entirely clear when dioxin was discovered. U.S. congressional reports claim that dioxin was first identified as a compound in the mid-1950s when synthesized by a research technician, but other documents suggest that scientists have known of dioxin's existence for much longer. A "Dioxin Timeline" published by the Environmental Working Group states, for example, that scientists have been "aware of dioxin's toxicity" since the late nineteenth century, when workers in a German chemical factory began developing severe rashes on their faces as a result of exposure to the substance.[13]

As Payne's statement suggests, the term "dioxin" has been familiar to the general public for decades, thanks to a series of locally based, but often globally connected, environmental disasters. Most notably, 2,3,7,8-TCDD was present in Agent Orange, a chemical defoliant used by U.S. troops during the Vietnam War that has since been traced to myriad birth defects and cancers. In the North American context, 2,3,7,8-TCDD was the primary toxin at stake in the 1970s-era Love Canal crisis, which concerned a community in the Niagara Falls area of New York State that had suffered from decades of chemical exposure. It was also released at high concentration around Times Beach, Missouri, in 1982, when much of the town's streets were sprayed with motor oil containing the toxic substance.[14] The word "dioxin" resurfaced in North American news in the late 1990s, when Karen Houppert published an article in the *Village Voice* warning of a "likely carcinogen" called dioxin in tampons, due to the bleaching process of materials involved. Houppert's article inspired various antitampon movements in the United States and abroad and may have had a role in encouraging major producers to change their manufacturing practices in an effort to limit dioxin production.[15]

Dioxin has also been a cause of serious concern elsewhere in the world. Japan, for example, experienced high dioxin emissions for much of the later twentieth century, due to a decades-long use of incinerators as the primary means of waste disposal throughout the country. In 1999, the Japanese government introduced policy to prevent the release of dioxin and PCBs into the environment, and studies show that Japanese dioxin emissions have indeed been drastically reduced.[16] In many areas of western Europe, dioxin has been identified at extreme levels in animal-derived food sources, even quite recently. To cite a few examples: international warnings were issued in 1999 regarding poultry and eggs originating from Belgium; in 2008 regarding pork products from Ireland; again in 2008 regarding the milk used to make *mozzarella di bufala* in Italy's Campania region; and in late 2010 regarding various animal meats and eggs from Germany.[17]

The Agency for Toxic Substances and Disease Registry reports that "for the general population, more than 90% of the daily intake of CDDs, CDFs, and other dioxin-like compounds comes from food, primarily meat, dairy products, and fish," but also, to a much lesser extent, fruits and vegetables.[18] Once it enters a body, dioxin generally settles in the liver and fatty tissue, and it is not entirely clear how long it takes to leave, mostly through stool but also through urine and breast milk. In the case of 2,3,7,8-TCDD, the average time it takes for half of the dioxin in a body to go away "is highly variable and may take from 7 to 12 years."[19]

Also somewhat nebulous is just what dioxin might do to a body, although ample evidence suggests that it is not good. Much of the available data on dioxin exposure has come from people exposed to very high levels of 2,3,7,8-TCDD in Seveso and Times Beach. The most immediate and obvious result of direct exposure in Seveso was chloracne, a skin condition that results in lesions or very extreme acne, can take years to go away and leaves significant visible scars. Some readers might recall images of the former Ukrainian president Viktor Yuschenko, who is said to have been intentionally poisoned by dioxin in 2004 and still bears the chloracne scars on his face to prove it. Additional human results officially traced to high levels of dioxin exposure include other rashes and skin conditions, liver damage, and an elevated risk for diabetes. Most sources, including the ATSDR, World Health Organization, and U.S. Department of Health and Human Services, now acknowledge that dioxin can also cause cancer. The U.S. Environmental Protection Agency classified dioxin as a known animal carcinogen and probable human carcinogen in 1985, and the International Agency for Research on Cancer declared dioxin a known human carcinogen in 1997.

Two years later, in 1999, the United Nations Environment Program (UNEP) warned that dioxin was a "concern" for all countries. It was at this time that the UNEP first drafted the treaty leading to the Stockholm Convention on Persistent Organic Pollutants. Written in 2001, finally in effect in 2004, and amended multiple times since, the Stockholm Convention is a global treaty dedicated to eliminating and reducing the production of harmful chemicals known to remain in the environment and bioaccumulate in bodily tissue.[20] The treaty does not ban the production of dioxin, but it does require participating parties to drastically reduce their dioxin emissions. Notably, neither Italy nor the United States has ever ratified or enforced it.

As is the case with so many toxins, the majority of dioxin research has been conducted on animals. Here the results are more concrete. Effects of extreme dioxin exposure on nonhuman mammals include reproductive damage, birth defects, and reduced fertility; altered levels of sex hormones and reduced production of sperm; and cancers of the liver, thyroid, and more. The ATSDR states: "Animal studies suggest that the most sensitive effects (effects that will occur at the lowest doses) are immune, endocrine, and developmental effects. It is reasonable to assume that these will also be the most sensitive effects in humans."[21] As I discuss in subsequent chapters, the threat of developmental effects from dioxin has been one of the greatest sources of exposure-related fear for residents in both Seveso and Taranto.

So much of the available literature cites Seveso as one of the most-extreme-known cases of dioxin exposure, as well as one of the few that have led to scientific studies on humans. Addressed in greater detail in chapter 1, these studies point to increased incidences of breast cancer, thyroid malfunction, and genetic abnormalities in the region. Significant data is now also available regarding Taranto, where residents experience elevated rates of mortality attributed to a variety of cancers, as well as serious respiratory, digestive, and perinatal disease, when compared to the national population. Residents of both communities are increasingly aware of the risks they either did or still do face, thanks to ever-deepening cultures of information in both sites. And yet the widespread presence of dioxin and other toxic substances in vegetation, animals, and their own human bodies has been and is still hard to concretely perceive, due to dioxin's lack of notable physical presence, as well as the limited availability and extreme cost of chemical analysis.

In short, dioxin presents a sobering albeit nebulous reality for all, one that we can strive to limit but cannot entirely avoid. I highlight it in this book for its dominant role in Seveso and Taranto, and for the ways in which those two temporally distinct and localized Italian environmental crises are in dialogue with one another. I am also interested, however, in the ways in which dioxin engages questions of the universal, on scales both geographic and temporal. Not only is dioxin present worldwide, affecting all of us to varying degrees, but its ability to pass through the nonhuman environment and eventual food chains, bioaccumulating and biomagnifying, underlines the deep physical connection between all living beings and the world surrounding us—a world that is inevitably also in us. Like so many substances, dioxin may enter our bodies without our awareness of it doing so, through the air we breathe, the water we drink, the food we eat, and even the soil we touch. In this it is a reminder of how deeply we are embedded in our surrounding environment, a reminder that we may just be the sum of our parts but that our parts are many and changing, and not entirely within our control.

Dioxin encourages a reckoning with the transmutability of the lived world, the ability of so much and so many to change from one state to another, especially due to contact with another substance or being. Even more than that, it encourages a reckoning with the idea of transcorporeality, which Stacy Alaimo has extensively theorized as the movement of substances between and across bodies and matter. Writing of "interchanges and interconnections between various bodily natures," including chemical agents and ecological systems, and the "material interconnections of human corporeality with the more-than-human world," Alaimo underlines the way in which our bodies are never in stasis, but instead might be thought of as engaged in an ongoing process of embodiment.[22] Her emphasis on process is crucial not only to a general understanding of bodies, which are indeed always in flux, always adapting to and incorporating environmental stimuli, but also and in particular to an understanding of bodies and—and with—dioxin.

Concentrated exposure to dioxin, like other persistent organic pollutants, challenges subjects to acknowledge that the makeup of our bodies is ever-fluid, not entirely within our control, and not even entirely human matter. As Jane Bennett writes in her work on bodies and, in, and with the more-than-human world, "my 'own' body is material, and yet this vital materiality is not fully or exclusively human. My flesh is populated and constituted by different swarms of foreigners."[23] Thinking through concentrated dioxin exposure forces engagement with this "'alien' quality of our own flesh." It underlines that we cannot entirely know just how our bodies are changing, nor when or at what speed, and forces us to recognize the "animate transgressions" of which Mel Chen writes, when "proper intimacies . . . between human and nonhuman things" are blurred.[24] Wide-ranging and diverse, the theories of Alaimo, Bennett, and Chen, which inform my work here, are all grounded in a feminist new materialism particularly attentive to the importance of "lived experience, corporeal practice, and biological substance."[25] As I explore in the coming chapters, transcorporeal flows of toxic substances cannot be disentangled from questions of sex and gender, or from the ways in which female bodies and subjectivities are so very much at stake in the debates over health, reproduction, and caretaking that have emerged in both Seveso and Taranto.

On a conceptual level, thinking of our own bodies as deeply enmeshed in an ongoing flow of matter throughout the lived world is quite intriguing. We need only turn to the recent boom in mainstream public discourse about the human microbiome for evidence that people are increasingly curious about the interaction of our bodies with the surrounding world. The microbiome is the collection of bacteria, fungi, and viruses that contribute to our physical beings, and that continue to ebb and flow throughout our lives depending on the other things and beings with which or whom we come into contact. The National Institute for Health's Human Microbiome Project, established in 2008 to aid in understanding the possible role of microbes to treat disease, is one example of growing interest in this swarming collection of living matter.[26] Michael Pollan's 2013 *New York Times Magazine* article "Some of My Best Friends Are Germs" is another. Here the writer describes his own test results from the American Gut Project, a "citizen science initiative," whose work allowed him to start thinking of himself "in the first-person plural," an open vessel housing countless bacterial species.[27] These examples, along with myriad others and a growing industry dedicated to "gut health," all explore the beneficial potential of bacteria. In this they offer a certain amount of solace and regained agency in the face of rising rates of autoimmune disease and cancers worldwide: we might be able, they suggest, to purposefully introduce naturally occurring outside agents to our bodies in an effort to strengthen them.

Underlying this growing interest (and fueling what is only the newest branch of an ever-booming wellness industry) is the fact that our bodies are as permeable to harmful agents as they are to helpful ones. Whereas we can choose, for example, to eat fermented foods every day, adding what has been deemed healthy bacteria to our microbiome, we have much less control, as addressed above, over the various environmental pollutants that may also be shaping our (well-)being. An awareness of such susceptibility to harmful matter, of the possibility for our bodies to continue changing in unknown and frightening ways, can be quite disempowering, to use again the language of agency, which can in turn lead to a troubling sense of lack of control, not only of our surrounding environment but also of our most intimate selves.

This, I argue, is where additional tools for coping, comprehending, and regaining a sense of agency become so important. Proactive physical steps toward healthy bodies and environments are certainly of great importance, but they cannot be achieved without the emotional and cognitive comprehensions that lead the way. Such comprehensions might indeed be our greatest resources for a continued existence in and with our bodies and the world. This is where narrative has so much to offer, serving as a resource for greater understanding via increased information, imagined experience, empathic response, and possible action. I argue here that dioxin—and, more universally, the process of toxic embodiment—necessitates narration so that subjects might affirm their own existence through creative engagement with the shifting bounds of the material world.

Moreover, dioxin necessitates narration—and the sense-making powers embedded in the processes of both telling and responding to story—so that subjects might maintain agency over their very own beings as *they* shift in unknown and perhaps unprecedented ways. While in the cases of Seveso and Taranto dioxin prompts narrative expression from others, it also tells stories itself, tracing histories and making physical connections between places and beings. As such, while much of the following analysis is focused on human experience and bodily health, dioxin's omnipresence necessarily pushes Seveso and Taranto beyond what Jeffrey Cohen has called "the diminutive boundedness of merely human tales."[28] The many stories stemming from and about these two places and their resident communities are necessarily stories of land and matter as much as they are stories of sentient beings.

As practitioners of the environmental humanities have fruitfully explored in recent decades, narration is a powerful tool for coming to terms with environmental change, advocating for environmental justice, and modeling new modes of coexistence. An exploration of serious challenges to health, land, and community in story form clearly offers sense-making space for narrators themselves. As scholars such as Marco Armiero (2014) and Ann Jurecic (2012), preceded by Sandra Steingraber (1997), Svetlana Alexievich (1997), Arthur Frank (1995), and others, examine, in narrativizing what has happened, what might happen, and what is still happening in the worlds both around and *within* them, narrators gain the opportunity for increased knowledge and comprehension. These "wounded storytellers," to use Frank's term, are able to trace out a progression of events, those events' eventual impact, and their own emotional and physical responses, as well as the responses of others.

Joseph Dumit (2012) writes that this is especially the case when it comes to narrating subjects naming their own at times nebulous symptoms rather than relying on messaging from corporations and governments.[29] By telling their stories, they become cognizant authors of their own corporeal processes even as they move in directions that cannot entirely be controlled, thus creating thoughtful catalogues of conditions and experiences that may otherwise remain unwritten. At the same time, by sharing stories of exposure and illness, narrators are able to situate themselves in relationship to other things and beings. Those who have shared their own stories of environmental illness, such as Alexievich's interview subjects or Steingraber herself, trace the ways in which the farm or factory down the road directly impacts the intimate lives of individual residents, as well as whole communities. In this, they draw out a vast web of environmental and ontological relationship.

I am attentive to the results of narrative for those who have told or are still telling their own firsthand stories of Seveso and Taranto. As such, I consider them throughout the chapters to come, alongside the work of those artists and authors who instead serve as witnesses, recounting the stories of others. I am even more concerned, however, with the potential effects of narrative as experienced by the vastly larger communities comprised by receptive audiences. In reading, watching, physically traversing, or otherwise experiencing a text, we all have the possibility not only to gain rational knowledge from the information shared but also to gain a deeper empathic understanding through imagined experience. While this is often a solitary venture, it does provide the chance to take part in what Arjun Appadurai calls a "'community of sentiment,' . . . a group that begins to imagine and feel things together," however disparately.[30]

Furthermore, as narrative texts allow for greater understanding of events and increased feelings of connection to both people and place, they facilitate potential action. As Serenella Iovino notes, "narrative representations are an essential instrument of action and knowledge: by re-framing and re-creating an event in its material-discursive patterns, they provide a necessary reconfiguration of meanings, of inter-active dynamics, and of ethical responsiveness, thus enabling constructive visions of the future."[31] Just as narratives can make events more real (often, conversely, through imagination), they can also allow receptive audiences to more clearly position ourselves in relationship to those or similar events, while informing our processes of contemplation, understanding, and problem solving regarding future actions. To be clear, when I write here of "receptive audiences," I simply mean readers and viewers who are "willing to consider or accept new suggestions and ideas."[32] What follows is *not* a work of reception studies, which might be based in analysis of viewer surveys or even use neuroimaging to assess subjects' brain function as they read. Instead, it considers a given text's potential to offer meaningful experience and increased understanding to audiences who are willing to engage, whether or not they have previous familiarity with the subject matter.

In this light, I am particularly inspired by recent work in an ecologically focused strain of narratology hinging on the notion of the storyworld. The cognitive narratologist David Herman describes storyworlds as "the worlds (= configurations of participants, objects, and places and sequences of states, events, and actions) evoked by cues contained in narratives," that is, the realms to which we are transported (and which we are encouraged to mentally construct) by dynamic texts.[33] Utilizing this notion, Erin James makes the case for what she labels econarratology: "Econarratology studies the storyworlds that readers simulate and transport themselves to when reading narratives, the correlations between such textual, imaginative worlds, and the physical, extratextual world, and the potential of the reading process to foster an awareness and understanding for different environmental imaginations and experiences."[34] As James and others argue, by experiencing storyworlds through text readers are able to gain greater awareness of and empathy for diverse environmental realities, thus taking knowledge gained within the reading experience back out to the lived world and allowing it to affect their subsequent behaviors. In the following chapters, I adapt James's approach to explore how acts of reading, as well as those of viewing and participating in physically grounded modes of reception such as walking, foster awareness and understanding for different environmental experiences but also for different corporeal ones. For as Seveso, Taranto, and their many narratives explore, the realms of nonhuman environment and human bodies can never truly be separated, each one shaping the other.

Recently, Corrine Donly has built on the notion of econarratology to propose the complementary "eco-narrative." Reworking a term broadly defined by Ursula Heise in the *Routledge Encyclopedia of Narrative Theory* (2005) as a category of "narratives about nature," including "mythological creation stories, science fiction novels and filmed nature documentaries," among others, Donly offers a more narrow definition of eco-narrative focused on the storytelling process.[35] Arguing against traditional conflict/climax/resolution plot models, they propose that storytellers might instead take inspiration from the nonhuman world to craft more playful and less anthropocentric narratives, thus working to "convey ecosystem" rather than resolve conflict. With this Donly proposes "a more ecologically responsible framework through which storytellers can enter into their storytelling practices," by reflecting the ways of animals, plants, and other sorts of matter through theme and plot, but also narrative structure.[36] In the realm of poetry, Donly's eco-narrative holds affinities with the sort of ecopoetics explored by Scott Knickerbocker, via texts in which "the most meaningful contact with nature occurs through form."[37] Thus in kinship with but more formalist than James's econarratology, Donly's eco-narrative sheds light on the ways in which a text can reflect the nonlinearity, playfulness, rhythmicity, relentlessness, and more in the nonhuman world—whether from the currents of a river, the behavior of animals, or the unpredictable life cycles of persistent organic pollutants. Both concepts inform my critical approach in this book.

As explained above, in my attention to the storyworlds of Seveso and Taranto's dioxin-tinged texts, I am particularly focused on what narrative immersion offers to those experiencing the narratives: how taking part in story as a reader or viewer allows subjects to make sense of new and nebulous realities (often, as Marco Caracciolo has argued, through cognitive identification with characters).[38] And yet the story, whether crafted by authors, filmmakers, or dioxin itself, remains a crucial part of the formula, especially in light of how flows and patterns of the more-than-human world might influence a narrative. I argue in the coming chapters that narrative helps audiences understand dioxin and its transcorporeal abilities through storyworld experience, which hinges heavily on content and the sense of knowing that it may foster. I also argue that it does so through a story's shape, what Herman calls "story logic"—taking on forms that are less than directly linear and often, as Donly proposes, not rooted in the traditional narrative arc of conflict resolution.

In emphasizing the power of narrative and the agency of matter in relationship to Italian texts and places, I am delightfully far from alone. Instead, the current study joins a rich and ever-growing body of work in the Italianist environmental humanities. In recent years, books by Serenella Iovino (*Ecocriticism and Italy*, 2016), Elena Past (*Italian Ecocinema beyond the Human*, 2019), and Enrico Cesaretti (*Elemental Narratives: Reading Environmental Entanglements in Modern Italy*, 2020) have helped shape a growing discourse around the relevance of Italian studies to the broader environmental humanities, just as they have ushered an ecocritical gaze into the field of Italian studies. Further contributions can be traced to that trio's edited volume *Landscapes, Natures, Ecologies: Italy and the Environmental Humanities* (2018), as well as Pasquale Verdicchio's edited volume *Ecocritical Approaches to Italian Culture and Literature: The Denatured Wild* (2016), and my own previous work *Landscapes in Between: Environmental Change in Modern Italian Literature Film* (2015). The aforementioned texts, along with recent articles in journals such as *Ecozon@* and *ISLE*, as well as a special eco-media cluster in *The Italianist* film issue, all contribute to what Verdicchio describes as an expansion of the "terms by which an Italian ecocriticism might reveal a . . . rich environmental literary and cultural tradition."[39] More often than not, these diverse studies and critical interlocutors are united in their attention to the particularities of Italian environmental history, the notion of landscape as text and the potential for creative expression to serve as a form of both education and resistance.

Toxic Matters is most overtly in dialogue with the first three titles listed above, due to their nuanced attention to what Cesaretti calls "the narrative eloquence and expressive energy of materials," and Iovino, "storied matter."[40] In the coming chapters, my work looks particularly toward that of Iovino and Past for their shared attention to environmental toxicity, the benefits of moving slowly, and a certain measure of personal situatedness. It also celebrates the liberatory capacity of narrative (Iovino) and the use of first-person interview and walking as meditative methodology (Past). In short, the present study declaredly takes inspiration from and converses with these scholars and their texts. At the same time, it seeks to further broaden the evolving discourse of an Italianist environmental humanities, through its attention to econarratology and the bright light it shines on two significant Italian case studies and their unique narrative output.

The stories that trace dioxin in Seveso and Taranto span realist and speculative modes, first-person testimonies and fictional reelaborations, literature and cinema, rhythmic orality and scientific legalese. They take on a vast range of forms and genres, from autobiographical narratives written by factory workers in both sites to novels such as Laura Conti's *Una lepre con la faccia di bambina* (1978, A hare with the face of a child) and Cristina Zagaria's *Veleno* (2013, Poison); from oral interviews housed in the Legambiente archives, to stories told in court documents and newspaper articles; from the horror film *Nightmare City* (Umberto Lenzi, 1980) to the experimental documentary *Non perdono* (Grace Zanotto and Roberto Marsella, 2016, I do not forgive / Nonforgiveness); and from a "memory path" in Seveso's Oak Forest to a performative act of hemp planting by the artist Noel Gazzano. When analyzed in conjunction with one another, these diverse primary texts allow us to reevaluate and to privilege the role of storytelling devices in making sense of ecological and human health disasters. At the same time, they allow us to interrogate the mutually constitutive intersections between biological limits and the creative capacities of people to think through and against the larger forces that may do us harm.

The following chapters are written for readers who may or may not already be familiar with Seveso, Taranto, and their respective crises. My intention is to provide enough background information that a wide readership may follow my discussion of the narratives produced in response to the crises, without getting so bogged down in albeit important auxiliary detail as to lose the primary thread. At the same time, I hope to provide enough descriptive summary that the narratives discussed, most of which have yet to appear in English translation, may still be accessible to readers who do not understand Italian and thus might not otherwise engage with them. The "we" whom I address in this book, my own intended audience, includes scholars and students, lay readers and activists, residents of the places featured herein, and those who may never touch Italian soil. I do not aspire for this to be an exhaustive case study of either place, community, or narrative output but, rather, a critical consideration of one particular and powerful result of the significant dioxin in Seveso and Taranto: an equally significant storytelling practice. In so doing, I hope to underscore the transformative potential of narrative, while fostering in readers a deeper understanding of industrial pollution and resultant eco-corporeal crisis—contemporary realities that the universal "we" can no longer afford to ignore.

Chapter 1 introduces the Seveso disaster by offering a clear overview of the disaster itself, as well as the days and months following in its wake. Engaging with first-person narratives, archival research, and the work of the sociologist Laura Centemeri and others, it underlines the lack of clear communication between state and factory officials and residents, and the turmoil this caused. Chapter 2 then explores a narrative treatment of the disaster in Laura Conti's writing, particularly her aforementioned short novel *Una lepre con la faccia di bambina*. I argue here that Conti was ahead of her time in explicitly offering a fictional text as a means to aid comprehension of a complicated new eco-corporeal reality. In closing, I consider Gianni Serra's 1988 adaptation of the novel for television, and the ways in which he uses that visual medium to further Conti's exploration of how imaginative practices might help sense-making processes.

Chapter 3 introduces readers to Taranto and Ilva, from the city's foundation to the steelworks' creation and on to the present day. Following the model established above, the subsequent chapter, 4, examines a set of narrative texts set in present-day Taranto. Here I emphasize a common attention to spatial mapping, as well as the urgent plurivocality of some of the narratives that have emerged in response to Taranto's ongoing crisis. I argue that while articulating the city's geographical coordinates serves to draw readers into storyworld experience, the myriad voices represented work to dismantle a singular authoritative narrative about Ilva and its effects. Chapter 5 remains with Taranto but focuses on cinematic expressions, primarily in the form of documentary. I argue that the films in question communicate a common urgency to spread factual knowledge about a crisis still underway, just as they explore unique narrative conceits. In conclusion, I challenge mainstream fiction films to follow suit.

The book's final chapter, 6, puts Seveso and Taranto in direct dialogue with each other by considering the landscapes of each location as texts unto themselves. It begins by analyzing present-day Seveso's Oak Forest, cultivated in the site of the former ICMESA factory, then moves on to consider a series of performative and collaborative actions undertaken in present-day Taranto. In both cases, I consider the possibilities of bioremediation and advocate for turning to the land itself as a site of historical memory and storytelling. Between chapters, I include excerpts from first-person interviews I conducted in Seveso and Taranto during the summer months of 2015, 2016, and 2018. Digitally recorded on-site and then later transcribed, these interviews often took shape as conversations shared over coffee, meals, and long, ambling walks. The interviews lend a necessary plurivocality to this study, in which I argue that official narratives must be both complicated and complemented by a multitude of voices. Comprising a participatory ethnography, they feature some of the residents, activists, and artists who are shaping Seveso and Taranto's lived stories as they unfold.

CHAPTER 1

Seveso
Making Sense

In 1976, the northern Italian town of Seveso had a population of approximately seventeen thousand people. It boasted a long-standing tradition of artisanal furniture craftsmanship and the possibility for a quiet suburban life. Many residents had workshops nearby and kept small gardens and chicken coops in their backyards. The local economy had diversified slightly three decades prior, in 1945, when the ICMESA factory opened just over Seveso's border in the neighboring town of Meda. ICMESA (Industrie Chimiche Meda Società Azionaria) was a subsidiary of the Swiss fragrance and flavor manufacturer Givaudan S.A., acquired by the multinational Hoffman La Roche group in 1963. ICMESA was one of many manufacturers to establish a northern Italian industrial presence in the wake of World War II. Thirty-one years after the factory's opening, it maintained 170 local employees, many of whom had migrated to the area—first from the Veneto region in the 1950s, then southern Italy in the 1960s—and now lived quite close to factory grounds.

These employees worked to produce various essential oils and essences for use in perfumes and cosmetics. Since 1970 they had also produced trichlorophenol, a synthetic compound used in pesticides and herbicides, including the toxic defoliant known as Agent Orange and utilized by U.S. troops for tactical deforestation during the Vietnam War. On Saturday, July 10, 1976, ICMESA's trichlorophenol reactor overheated due to a series of small missteps made in the course of shutting down operations for the weekend, standard practice at the time. This led to a buildup of excessive pressure in the reactor, causing a relief valve to open and release a massive amount of the by-product 2,3,7,8-tetrachlorodibenzo-p-dioxin in a large cloud held together mostly by sodium bicarbonate.

Due to strong winds that day, the cloud moved southeastward from ICMESA, across Seveso and into the towns of Cesano Maderno and Desio, in a linear path stretching about six kilometers. The resultant dioxin contamination was not as concentrated but significantly more widespread than it would have been on a less windy day. In all, approximately fifteen square kilometers of land were exposed to dioxin that afternoon. As is often the case in industrial environmental disasters, ICMESA workers and their families—many of them still relatively recent migrants—were disproportionately affected by immediate exposure to the toxin because of residential proximity. Many thousands of other area residents, however, were also exposed. According to reports released in 1983, 54 percent of Seveso, 52 percent of Cesano Maderno, 20 percent of Meda, and 18 percent of Desio were contaminated by dioxin.[1] The incident was quickly labeled the Seveso disaster and is often referred to simply as "Seveso," due to ICMESA's location alongside Seveso town limits, and the extreme contamination experienced there.

Although soil samples were eventually taken in all of the towns listed above, it is hard to know the precise amount of dioxin released into the sky, soil, and eventual food chain from the explosion. Dioxin analysis is a costly process even now, and, as I explain further below, factory and local authorities were slow to evaluate the extent of the disaster. In a 1991 study on Seveso's aftermath, Bertazzi et al. estimate the amount of TCDD released on July 10, 1976, to be 34 kilograms or higher,[2] whereas others cite instead a range of 300 grams to 130 kilograms.[3] For comparison, Thomas and Spiro report that in 1989, "total annual US air emissions of polychlorinated dibenzo-p-dioxins and polychlorinated dibenzo-furans (PCDD/Fs) from all known sources are estimated to be about 400 kilograms."[4] A total of 400 kilograms throughout all of North America over the course of one year suggests that 34+ kilograms in one isolated site at one isolated moment was a great deal indeed. To date, the Seveso disaster marks one of the greatest-known immediate releases of dioxin into the atmosphere worldwide, as well as one of the greatest cases of human and animal exposure. A 1998 study found, for example, that, upon eventual analysis "some individuals in the exposed population had among the highest serum dioxin levels ever reported."[5]

While Seveso is internationally notable for the extreme amount of dioxin released all at once, the case also stands as a negative example of mismanagement regarding not just safety operations but, even more significantly, communication with the affected communities. Perhaps most remarkable in accounts of the disaster are repeated chronicles of delayed and limited public notice, coupled with mass uncertainty as to what might come of residents' immediate surroundings and bodily health. Regional authorities were alerted to the accident only the day after it occurred, when the local ICMESA manager, Paolo Paoletti, briefly explained to Seveso's mayor, Francesco Rocca, that the trichlorophenol reactor had blown. In his Seveso memoir, *I giorni della diossina* (2006, The days of dioxin), Rocca recalls Paoletti saying: "Well, TCF is what is produced. But what came out we can't know, because at higher temperatures the substance can create other products. Nearby families ought to be warned not to eat fruits and vegetables."[6] Paoletti's response represents ICMESA's general approach to Seveso recovery efforts: admitted uncertainty about what had in fact been released, along with a slightly more pointed warning to be cautious about interacting too closely with the local nonhuman environment. What Paoletti's statement shows to be so unsettling about the Seveso case is a combination of limited knowledge as to just what was seeping into the environment and its inhabitants, and a strong suspicion that it could not be good.

Over the years, Seveso scholars have compiled various timelines of the events following the ICMESA explosion. From the investigative writer John G. Fuller's engaging but occasionally sensationalist *The Poison That Fell from the Sky* (1977) to the environmental sociologist Laura Centemeri's heavily researched *Ritorno a Seveso* (2006, Return to Seveso), multiple book-length studies seek to make sense of just what happened in Seveso not only on but also after July 10, 1976. I add to this list Rocca's aforementioned book, as well as Laura Conti's *Visto da Seveso* (1977, Seen from Seveso), Daniele Bacchessi's *La fabbrica dei profumi: La verità su Seveso, l'Icmesa, la diossina* (1995, The perfume factory: The truth about Seveso, Icmesa, and dioxin), and Massimiliano Fratter's *Seveso: Memorie da sotto il bosco* (2006, Seveso: Memories from under the woods). A reading of these works—alongside the significantly larger number of scientific studies published in academic journals and myriad articles on websites and blogs—reveals frequent discrepancies regarding dates (of key discussions among officials, for example) and amounts (of people, land, and dioxin). These discrepancies point to a lack of consistent communication and documentation in Seveso's aftermath. What the aforementioned texts do universally project is a portrait of delayed public notice, pervasive uncertainty, and general mistrust of authority among the local community. To quote Fratter, "The lack of clarity and the contrasts that played out among the various authorities were clearly perceived by the community, which was not able in the moment of the emergency to find, with promptness and clarity, the responses that were necessary in the face of an 'invisible' poison."[7]

I offer here an abbreviated timeline of official communication, based on documents authored by the Health Office of the Greater Lombardy region (Regione Lombardia Assessorato alla sanità, Giunta regionale) and now housed in the Fondazione Micheletti archives in Brescia.[8] This sequencing of events underscores an apparent lack of urgency regarding public information in Seveso's wake:

July 10—An ICMESA foreman phones a production manager to come and inspect the factory. The manager phones the local health officer (*ufficiale sanitario*) for Seveso and Meda, as well as Meda police, to report the incident.

July 12—ICMESA officials contact police in both Seveso and Meda and tell them that residents should avoid eating fruits and vegetables from their gardens.

July 13—The local health office issues a notice to area police and mayors' offices stating that they do not yet understand what happened at ICMESA or what substances were released in the explosion.

July 15—The local health office sends a letter to regional health officials, police, and mayors' offices reporting the deaths of area animals and warning of probable diffuse toxicity. Seveso mayor Francesco Rocca issues his first public notice to avoid fruits, vegetables, and contact with soil.

July 16–19—Children are admitted to hospital with chloracne and gastrointestinal illness. ICMESA workers, who had returned to the factory on the Monday following the explosion, strike over growing concerns about onsite safety.

July 17—The first news articles about the explosion are printed, in the Milan-based *Il Giorno* and the Milanese section of the national *Corriere della Sera*. A more extensive article appears on the front page of that newspaper's national edition the following day.

July 18—After a site inspection, the director of the Milan Provincial Laboratory for Hygiene and Prevention, Professor Aldo Cavallaro, suspects that dioxin was released in the explosion alongside trichlorophenol, and orders analysis. The ICMESA factory is closed.

July 20—Dioxin is confirmed in soil samples.

As many Seveso chronicles note, top ICMESA/Givaudan officials resided in Switzerland and were not physically present for the explosion's immediate aftermath, thus further delaying processes of on-site analysis and communication to local communities. Centemeri writes that "dependence on information provided by Givaudan made it difficult for local authorities to know what line to follow" as they waited for information to be relayed from Switzerland before taking action.[9] Following the ten days it took to confirm the presence of dioxin, many letters and phone calls were exchanged between local authorities, regional health officers, and Givaudan executives. The contaminated areas of Seveso and neighboring communities were eventually divided into three zones based on estimated levels of exposure (Zone A, Zone B, and the "Zone of Respect"). On July 26, the first wave of residents from the heavily contaminated Zone A were evacuated and moved to a hotel on the outskirts of Milan. Within the week, a total of 736 residents had been evacuated. In Zone B, children and pregnant women were removed by bus during the daytime but allowed to return at night, and all residents were warned to follow a set of guidelines issued on August 13 by regional public health headquarters in Milan. Authored by that office's lead investigator, Vittorio Rivolta, the statement explains that residents are at risk of intoxication from contacting outdoor surfaces such as terraces and roofs, as well as from eating local vegetal or animal products. It also contains the following Addendum:

1) It is extremely important to wash hands immediately and thoroughly. . . . We also recommend a frequent and accurate cleaning of the whole body (bath or shower every day). . . .
2) Being known that subjects who have come into contact with the substance can experience skin reactions such as hypersensitivity to sunlight, it is recommended that you not expose yourself to the sun for prolonged periods.
3) The ingestion of contaminated foods provokes a diffusion of the substance in the body, with damages located primarily in the liver and kidneys. Because of this it is highly dangerous to ingest any sort of animal or vegetal products from the zones indicated above. It is thus

prohibited to cultivate or forage herbs, flowers, fruit, or vegetables in the polluted regions, and to raise animals other than domestic pets.
4) Any work that involves the movement of dirt is forbidden, as it necessitates contact with skin and the raising of dust. . . .
5) In order to avoid raising dust, vehicle speed on nonasphalted streets should not exceed 30 km per hour.
6) It is prudent that all people exposed to the risk of contamination abstain from procreation for a preventive period of at least six months.[10]

In the above, the nearly five thousand residents of Zone B are instructed to vigilantly cleanse their bodies; avoid exposing their skin to the sun; not eat the vegetables or animal products they have cultivated; not disturb the soil surrounding their homes; not drive at elevated speeds, and avoid procreation. In short, they are told that their normal existence has become dangerous, that something out there might have damaging effects not only to the surrounding nonhuman landscape but also to the bodies of the humans and animals in its midst. And yet beyond damage to liver and kidneys from ingestion, and possible hypersensitivity of skin to sun (the original Italian for point number 2 uses the verb *potere*, an expression of potential rather than certainty), residents are not explicitly told what will happen if they engage in the discouraged behaviors. What's more, the direct agent of these eventualities, dioxin, is never named in the addendum, thus further contributing to its evasive tone. As a result, many residents doubted the severity of the above and similar warnings, and, as time stretched on, they began to resent the imposition of what appeared to be unnecessary cautionary measures.

Oral testimonies collected by the local chapter of Legambiente, Italy's most prominent environmental nongovernmental organization, underline that this pervasive sense of public mistrust was fueled by dioxin's very nature. Able to impact even when invisible and according to an unpredictable timeline, the potent yet imperceptible substance held disconcertingly close ties to the realm of the hypothetical rather than the real. Resident Maria Luisa Sartori describes: "A strange sensation. I remember this well. If someone says 'I have a stomachache, I have a headache' it's a reality that you know. But this invisible thing . . . And this sensation that I'm carrying something around with me, in the sense that . . . not that . . . In the sense of something impalpable that can happen to you from one moment to the next. And so we were terrorized."[11] In its ability to cause change and, indeed, damage to living organisms, without an obvious physical form, dioxin poses a particular ontological challenge as it asks people to believe in something for which they have little overt evidence.

Particularly striking in Sartori's testimony is her reference to physical symptoms as a standard proof of presence lacking in the wake of dioxin exposure. In their work on the social function of symptoms, Joao Biehl and Amy Moran-Thomas write that symptoms can be "a necessary condition for us to articulate a relationship to the world and to others."[12] In naming a symptom we are able to identify a physical sensation that others will understand, thanks to the power of language. My headache may feel different than yours, but we agree upon the word used to describe that certain range of physical experience involving heads and pain. Naming a symptom places that experience in the realm of the collectively known. And yet a symptom is only a secondary effect of a primary thing: a headache is not a chemical but a result of that chemical triggering my body's physical sensors. Sartori's recollection underlines that, in cases of dioxin exposure, even that secondary effect of its presence can be out of reach, still to come at an undisclosed time. Seveso's dioxin thus forces engagement with possible futures in a moment of immediate crisis, as it pushes against traditional bounds of knowability—scary stuff indeed.

Centemeri reiterates that dioxin's unknown and largely unknowable status in Seveso would have required a particularly strong and unified language from authorities for citizens to have believed more firmly in its existence: "The 'invisible' risk of dioxin, in order to be made perceptible and present, should have imposed extremely rigorous, coherent, and realistic behaviors on the part of those who were seen as possessing knowledge, a knowledge that people inferred based on action."[13] Laura Conti notes, too, "the uncertainty regarding what to do was interpreted as an uncertainty regarding knowledge, and thus an uncertainty about the danger of the dioxin."[14] As we have long seen throughout global history, communities turn toward a projection of strong leadership in uncertain times, often a projection that is underlined by decisive action. Absent such leadership in the wake of Seveso, residents began to question just how much of a risk dioxin truly represented. I should note, as Conti does in her first-person testimony, that the scientists and medical experts who became involved in the Seveso aftermath were themselves working to make sense of great unknowns, and hesitant to cause undue alarm.

The explosion on July 10 did harm bodily health in immediate ways, as well as future ones. As remarked upon in so many testimonies, the first indication that something was wrong after the explosion came from area animals. In an interview conducted in 1999, resident Aldo C. recalls: "I didn't see any more swallows, and when you don't see swallows it's bad because something has really happened. And when the dioxin came out no one saw any more swallows, they had all vanished."[15] More than just swallows disappeared: in the following days and weeks thousands of small animals, both wild and domestic, died. Smaller animals, such as rabbits, cats, and chickens, were the most acutely affected, due to toxicity of liver, lungs, and kidneys. Larger animals also perished, however, and regional authorities eventually ordered the slaughter of close to eighty thousand animals, including cows and sheep, primarily so as to remove them and their products from the local food chain.

Within the course of one week from the disaster the first human health incident was recorded, when an infant was brought into an area hospital with a severe case of the skin condition chloracne. After this first case of chloracne, approximately two hundred more were reported, along with cases of gastrointestinal illness, eye irritations, and headaches.[16] Such physical manifestations—easily comprehended symptoms—indicate only the instant effects of dioxin exposure, without yet speaking to long-term effects now known to include reproductive and developmental problems, changes to hormonal and immune systems, and various cancers. A twenty-five-year study of Seveso mortality, conducted through the same Mangiagalli hospital in Milan where many patients were first treated, confirms remarkably excessive rates of mortality from lung cancer in men and diabetes mellitus in women, as well as other terminal illnesses and disruption of gene expression in both sexes.[17]

As time stretched on in the summer of 1976, feelings of fear and anxiety were most acute regarding the status not of present human bodies but of potential ones. While some people questioned the risk of eventual illness to adult bodies exposed, even more debated the risks to fetuses carried by women who were pregnant during or shortly after the explosion. Such debates were conducted not only by those in the local community but also, and perhaps especially, by those outside of it. At the time of the disaster, abortion was illegal in Italy under any circumstance but had become a hotly contested issue; it would become legal within the first trimester of a pregnancy two years later, in 1978. Special exemption was soon granted for women in the zones effected by Seveso, however, and the first three known post-Seveso abortions were performed in Milan on August 12, with more to follow. The total number of women who ultimately decided to end their pregnancies cannot be known, as some traveled to other countries for the procedure or used clandestine methods, in the wake of what became significant uproar around the issue.

Different groups contributed to the abortion debate: health officials and scientists who warned of possible birth defects from the teratogenic dioxin; large swaths of anti-abortion advocates, emboldened by the pope's August 15 statement that the women in question were being exposed to "psychological subjugation"; feminist groups eager to advocate for women's right to self-determination; and others who had been pushing for the legalization of abortion for years. In her 1990 study of the reporting surrounding the Seveso disaster, Barbara Mascherpa argues that heavy coverage in local and national newspapers only stoked the flames. She maintains, in particular, that special-interest groups with media ties "tried to instrumentalize the Seveso case, transforming it into an opportune occasion to strike their adversaries . . . and, at the same time, reinforce their own ideology and power over public opinion," leaving pregnant women "the victims of this game without scruples."[18] While not surprising given the heated national debate already surrounding abortion at large in 1970s Italy, this meant that the private health decisions of women already living in a state of turmoil became highly public in Seveso's traditional Catholic community. It also meant that Seveso became a "woman's issue" in the eyes of many, particularly those who resided outside of the community. This is not to say that it was seen as an issue for women to resolve or make informed decisions about but rather that it became a social issue regarding women's bodies as objects, potential vectors, and platform for public debate.

The "abortion question," as it came to be known, again speaks to the ontological complexities posed by dioxin. It stands as an overt indication of the struggle to reconcile shared moral values with the material and temporal ambiguities of the toxic substance. Acknowledging the TCDD's potential to lead to harmful future effects meant attributing temporally durative power and agency to a nebulous entity, just as it meant entering into difficult conversations about the weight of moral and religious stances in the face of a shifting environmental and human health reality. It pitted science and ideology against one another in a realm that again felt largely hypothetical, as those involved once more had to grapple with an extended futurity, imagining what might yet come.

As I discuss in greater detail in chapter 6, the decades since the Seveso disaster have seen slow but steady efforts to remediate a polluted nonhuman environment, understand long-term human health impacts, and address lingering feelings of resentment, doubt, and shame in the local community. Many of the residents with whom I have spoken during my visits to Seveso note the embarrassment they felt in the years following the disaster when they traveled outside of the region. Patrizia, ten years old at the time of the explosion, described a class field trip a few years after the incident, during which she and her classmates felt too ashamed to tell fellow train passengers where they were from. Anecdotally, Patrizia's fears seem to have had some merit. A friend of mine who grew up outside of Rome in the 1980s once shared the following: when she and her brothers were children they would occasionally wave their bare feet in each other's faces, to the collective retort, "Eww, Seveso, Seveso!"—the implication being that their feet smelled so bad they were toxic, just like that ill-fated city to the north.

The disaster's most notable result on a less intimately discursive scale, alongside significantly increased information about dioxin and its behaviors, was the adoption of new European Union legislation regarding industrial accidents. Known as the "Seveso Directive," Council Directive 82/501/EEC was first passed on June 24, 1982, then subsequently amended in 1996 (Directive 96/82/EC, Seveso-II) and again in 2012 (Directive 2012/18/EU, Seveso-III). The Seveso Directive "aims at the prevention of major accidents involving dangerous substances," while also seeking to limit "the consequences of such accidents not only for human health but also for the environment." It involves a series of requirements for industrial operators and member state authorities, which focus on "prevention, preparedness, and response." At the same time, the Directive outlines the rights of citizens in the wake of an industrial accident. Notably, it states that "the public concerned needs to be consulted and involved in the decision making for specific individual projects."[19] As I discuss in chapter 6, the residents of Seveso and surrounding communities were eventually involved in decision-making about what to do with the former ICMESA factory site—but it took quite a while to get there.

In tracing out the above sketch of the Seveso disaster, I wish to emphasize two things: a pervasive lack of clarity surrounding dioxin and the risks that it posed, and a general lack of agency and breakdown of communication among members of the affected community, as encouraged by inconsistent official communication and the very nature of dioxin itself. These are just two aspects of a deeply complicated and multiaxial event. Also put into question by the event were the relationships (both environmental and legislative) between the local and the transnational; the role of private industry in producing a substance related elsewhere in the world to chemical warfare; the impact of national media and political debate on a local crisis; women's right to self-determination; an already decades-long history of odd smells and occasional animal deaths stemming from the ICMESA factory; and more. I focus in particular on the realms of knowledge, communication, and affect (uncertainty, fear, disempowerment) in the face of dioxin exposure at Seveso, as these are the realms most heavily explored in creative narrative response to the disaster—the realms that storyworld immersion so powerfully allows audiences to experience and sort through. In the next chapter, I consider immediate and fairly local narrative response to Seveso as represented by Laura Conti's writing, especially the short novel *Una lepre con la faccia di bambina,* before then offering a brief survey of other narrative engagements.

FIRST PERSON

Members of the Circolo Legambiente Laura Conti Seveso

In the spring of 2015 I had the opportunity to meet with members of the Circolo Legambiente Laura Conti Seveso, a local chapter of Italy's most active environmental NGO. Founded in Seveso around 1990 and based in Italian feminist principles emphasizing the practice of relationships, the Circolo is involved in various local environmental and social initiatives. These include the rehabilitation of a local nature preserve (the Fosso del Ronchetto), the curation of the Ponte della Memoria history project, the maintenance of an archive dedicated to Laura Conti and publications surrounding the Seveso disaster, and contemporary antipollution campaigns. We met on a warm day on the grounds of the Villa Dho, a historic estate that also houses the Casa Aperta, a social services center and group home for young women. After a delicious communal lunch with some of the Casa Aperta residents, Circolo members and I retired to a meeting room where I asked the members to talk about their experiences with the Seveso disaster and its legacy. What follows are excerpts from that conversation.

> LELE: I'll begin. I was twenty-one at the time. I remember that day because the strong smell that regularly—well, let's say that we *knew* because it came from the factory fairly often—that Saturday morning it was much, much stronger. Now, remembering a smell is very hard because your memory loses this sensation. But if I had to use one word, I would say it was the smell of a pharmaceutical product, like that but also with another something to it.
>
> Since I was fairly young I reacted with curiosity, trying to understand this event. Also, more than anything, all of us—my family but also my friends and other young people who lived this

experience—we were all very worried. And so we expected that the town, the public officials, would give official responses. Unfortunately, this never happened because neither the scientists nor the researchers, nor the public officials, nor the church were able to tell us, to give us *definite reasons*. And this made the anxiety grow for the whole local population. Then seven hundred people were evacuated for about two years. We had this fear, we would say to each other, "But what will happen in five years?" Those years eventually did pass, and now forty years have passed, and as it turns out, a whole lot of people didn't actually die. It was more the huge worry, the anxiety.

MAURIZIO: I was twenty-five, and I remember that Saturday, in the evening, you could smell an odor, an odor that you couldn't pin on chemical substances, but an odor like sulfur, rotten eggs, that kind of thing. I should explain that where I lived—I had lived there for five years, since I was twenty—you would normally only smell the odors from the factory when the weather changed. So when the weather changed you could smell this odor that mostly I remember was the odor of perfumes, because for a few years the factory produced perfumes, I think. When we smelled these perfumes we would say, "Oh, the weather's changing—you can smell ICMESA!"

That evening, though, you could smell an odor that was not a perfume. And the weather didn't change because from the time of the accident to the first rains, I think two or three months passed. I remember around then the priest from San Pietro church, in August, he held masses to bring on rain because people said that the rain would help to break down the dioxin dust and contain it in the soil rather than let it spread. And so there was this propitiary "rain mass" that finally had its epilogue in September or October when there was a great deluge, a devastating rain that flooded Seveso, basically right where I lived. It's a memorable event because it only happened that one time in sixty years.

ANGELA: Do you know that according to Laura Conti this event reduced deaths? But it also distributed them because—since dioxin is democratic—it went all the way to the Adriatic, slowly, slowly, slowly . . . and so, in part we still have it in the land here, but the flood also distributed it, it's democratic. Beautiful that flood.

LELE: Another aspect that really changed the story regarding chemical accidents was the fact that these chemical factories—here in this area in addition to ICMESA there were other, larger factories that produced chemical products—these factories were also appreciated from a certain angle because they created a lot of work, they employed thousands of workers—factory workers, office employees—which produced a certain wealth.

The most difficult thing was this: that the worry about a chemical accident was all directed toward *inside* the factory. That is, if a certain thing was produced in one area, measures were taken to cover the workers, make them wear gloves or masks while they worked. But no one ever thought that this problem could extend beyond factory walls.

LELE: Well, first of all, I should say that we had little communication, in the sense that there were only two state-run channels on TV. There were no private channels, no internet, no cell phones. And so we had a hard time talking to each other. We might get together in our free time, a stop at the bar or something, and we'd talk a little . . . you could say that dioxin had entered into our daily routines, in this sense.

But then during those years public meetings were organized, where the community was invited, something that obviously doesn't happen now—times have changed. But back then we really needed those moments, and you could say that from this angle we sort of needed to know more about this affair, but the newspapers and so on, they spoke about it with too much of a journalistic tone, too removed from us. And it's for this reason that we felt a need to search out, to find a place in which scientific communication was more relatable. This is the nature of Legambiente.

MAURIZIO: In terms of what happened and how it changed our lives, in the very beginning the task was to construct other roads because the main ones were closed. After this, at the end of the day, there wasn't a huge perception of the danger. Why? Because we live in an area that is densely industrialized in the chemical sector. From Varedo . . . two kilometers to the south you could smell a strong odor every day, so for us that was pollution, it wasn't something that you couldn't smell, we thought we knew well what chemical production was about. Acna Montecatini provided jobs for many, but also deadly tumors—many Montecatini workers got bladder cancer. So the perception of risk of danger was very present, but only for those who worked inside. And because the people who went to work there smelled bad odors, had these things in their heads . . . The awareness, the perception of a danger from things you couldn't smell, you couldn't see . . . When someone said there was this danger, it wasn't . . . it wasn't in our consciousness. This thing developed slowly . . . so in the first years there was really, beyond the annoyance of the closed streets, the "limited areas," there wasn't a lot of awareness about this problem.

MAX: Yes, and then I'm thinking, listening to you, Mauri, also about another kind of normality. For me the road was closed, which means that I had to go to school in a new neighborhood. Physically, physically and psychologically, I rooted myself in a different neighborhood. So my life—and I'm not here to judge whether for better or worse—changed, you know? The normality of the soldiers, right by my house there was the station for the soldiers who patrolled Zone A. I went around, I went to school on my bike and I passed by their truck. . . . It was the perception of a different type of normality, you know, the normality of "the next town over" . . . which was for us a way, I guess I'm trying to say a way to survive, but maybe that word isn't exactly appropriate. But it was a way to live with a minimum of—what's the right word? Normality . . . although it wasn't entirely normal, though, was it?

GIORGIO: Lots of people say: "But why talk about it? It's done with now. We want the future, we want lovely things in our life but this thing is an ugly chapter of Seveso. Why do you still want to talk about it?" But some people after so many years, given the continued [outside] attention, have gotten a little bit interested.

And I think I also heard some about the health question because of my mom . . . knowing some older people, various people who have died from various types of tumors. You can't say it's because of the accident, but you do have to ask yourself, with so many people dying from more or less the same thing—tumors that develop very quickly—maybe it is because of that. We all know that they can develop, that there's a greater likelihood for those types of illnesses here than in other areas.

MAX: It makes me think about the inability of science on one hand to recognize the real limits of the era, and to be able to say about the past . . . I don't mean to be able to offer answers but in some way to say, really honestly, "I don't have the answers." Because Giorgio's right, you know: here when someone gets sick you go there, everyone thinks about '76. The science of that era wasn't able to give answers, and still today science hasn't found a way to express what happened. I had the chance once to speak with Professor Moncarelli, who was the scientist who studied the accident most thoroughly. And he said, "Well, the answers have been published in *Science* [the journal]." Sure. But I dare you, first, to try to even find that issue, because it's hard to even find . . . and then! [We need to use] a language that doesn't trivialize the problem but also doesn't make it accessible only to a few erudite readers. This is a step forward. I have various friends whom I speak with often about this need to find a language that achieves, how can I say, an understanding between the doctor and the patient. Because otherwise you're in trouble. And this is what I think was missing. It was missing then, and it is still missing now. Like Giorgio was saying, there's this difficulty that you carry around with you . . . because I'm of the opinion that information should be shared, you know, with the appropriate explanations, of course. But this didn't happen, and the risk was minimized. It's true that it might have been (and in part was) only potential risk, but if you ask people about it, what they remember from that time is a fear of something unknown. And then we Westerners always need to have answers for everything—I mean all of us, you included! If we have a cold of course we need to get the pill that makes it go away, the doctor has to be some sort of magician who gives you the [cure]. There, the damage was invisible, you know. The kids got sick because . . . because why? Why some of them but not all? So this really added an element of confusion. We always use this example with the kids at school, that in Brianza a table is a table. If the table breaks you bring it to someone and they fix it. This is how we understand things: it's broken and then it's fixed, ill and then healed. And we still think like this, perhaps a bit less now because of experience, but some of this thought still lingers.

May 20, 2015

CHAPTER 2

Seveso Stories, or The Importance of Laura Conti

I have already mentioned Laura Conti in these pages, but I have not given her proper due. Born in Udine in 1921, Conti moved to Milan in the early 1940s and remained there until her death in 1993. A staunch antifascist, she participated in the youth resistance during World War II and was interned for ten months at a lager outside of Bolzano in northern Italy. Upon her release, she completed studies to become a medical doctor and was eventually appointed to Milan's Advisory Commission for Health and Ecology, also serving as a representative for the Italian Communist Party in the Lombardy region's legislative assembly. In both of these roles, Conti dedicated herself to educational and environmental causes, guided by what Stefania Barca describes as "an eco-Marxist vision of how social relationships always intrinsically include an ecological dimension."[1]

A dedicated scientist, educator, activist, and politician, Conti was also a prolific writer. Alongside numerous articles for local and national news outlets, her earlier publications include a novel based on her internment, *La condizione sperimentale* (1965, The experimental condition); a young adult novel about terminal illness, *Cecilia e le streghe* (1963, Cecilia and the witches); and a treatise on the importance of sex education, simply titled *Sesso e educazione* (1971, Sex and education). In the 1970s she wrote two books now fundamental to the Italian environmentalist canon that unite ecology, lucid prose, and a concern for workers' rights: *Il dominio sulla materia* (1973, The domination of matter) and *Che cos'è l'ecologia: Capitale, lavoro e ambiente* (1977, What is ecology: Capital, work and environment).[2] Her later publications grew even more ecologically oriented, with titles such as *Questo pianeta* (1983, This planet) and *Ambiente terra* (1988, Earth environment).

As Barca notes, Conti practiced a deeply "human-centered" ecology grounded in her attention to issues of class and labor, as well as her belief that "ecology had to do with the human body and its position inside the capitalistic power structure."[3] Her gaze was not limited, however, to a merely human sphere. Often described as an Italian Rachel Carson, Conti was very aware of the human capacity to destructively alter nonhuman nature, as well as the ways in which environmental change inevitably comes back to affect the health and well-being of living creatures. In an interview conducted just two years before her death, she claims that her work has never been ethically motivated but rather driven by feelings of "love" and "curiosity." She explains: "I study ecology because I am curious to understand how the mechanism of life functions. . . . I love animals, I love the human species, and I want it to last a long time."[4] Barca describes this as Conti's "commitment to social inclusion," a desire to both comprehend and advocate for all members of society.[5] When one reads her prolific oeuvre, it becomes clear that Conti has an expansive view of just who and what makes up a society—people, yes, but also animals and other forms of vibrant matter. Much like Carson, in her attention to the human species, Conti never loses site of our deep enmeshment with the nonhuman world—the material (and other) ways in which we and it are involved in an endless feedback loop. This perspective served her well when she turned her attention to Seveso.

As both an elected regional counselor and an environmental, social, and medical advocate, Conti took on an active role in sorting out the aftermath of the Seveso disaster. She was tirelessly involved in the legal proceedings and eventual policy changes that followed the initial event, due to her governmental appointments. As her published writing and private notes attest, she was a frequent participant in meetings and hearings about eventual remediation efforts toward both the nonhuman environment and human residents affected by the ICMESA explosion. Despite this significant level of participation, Conti still yearned for a more immediate engagement with the general population, in both Seveso and Italy at large. Her primary method of achieving this was through her writing, in newspaper and magazine articles but, even more significantly, in the personal reportage and social critique *Visto da Seveso* (1977, Seen from Seveso) and the short novel *Una lepre con la faccia di bambina* (1978, A hare with the face of a child).

Visto da Seveso is Conti's own account of the events surrounding the disaster. It begins, as do Fuller and Fratter's texts, by describing the day of the explosion before then tracing out the response of local health authorities, ICMESA officials, and area residents over the coming months. It stands as an extensive and carefully crafted record of the events following the chemical disaster, underlined by a strong desire to explain what Conti, like others, depicts as a pervasive sense of uncertainty. Throughout the text, which takes readers from July 1976 to April 1977, she addresses the need felt by officials to be cautious in their public communication; the resultant mistrust felt by the general public; the largely unknown nature of dioxin; and the largely unmet need for all involved to adopt a long-term temporal perspective. As her chronicle progresses, it is increasingly centered on the ongoing struggle of scientists, doctors, and various officials to determine and convey the actual toxicity of the dioxin released over Seveso on July 10: "All we could say was that those who lived in the poisoned zone had 'a certain probability' of taking into their bodies a substance that 'raised their probability' of becoming ill. This 'squared probability' was extremely hard not so much to explain intellectually, but to make enter into people's emotions. And so it sparked only the feeling of fear and thus the desire for people to distance themselves."[6]

Conti's tone can border on frustration in the text, a frustration with the results of ineffective communication. I read this as a testament to her deep engagement with both scientific discovery and human well-being: she is concerned for residents' health and wants them to have enough information that they take the risks of dioxin exposure seriously. In Conti's time, however, her tone was read as a dismissal of Seveso's residents as a close-minded and provincial people who chose to ignore warnings and, instead, take insult at their treatment by those outside of their community. A brief review of *Visto da Seveso* printed in *The Economist* on November 26, 1977, notes: "She becomes exasperated by what she feels is the pecuniary, isolationist and obscurantist attitude of the Seveso inhabitants. . . . [I]t does not seem to have occurred to Laura Conti that Seveso's inhabitants became outcasts because the average would-be helpers feared contamination by cancerogenous chemicals about as much as they would fear catching bubonic plague."[7] This review was, of course, published more than a year after the initial disaster. By then, considerably more was understood about the dangers of dioxin exposure, and news of Seveso's toxicity had had ample time to spread. An article published in the local *Il Corriere di Seveso* confirms this sentiment. The author writes of Laura Conti's snobbish detachment from the people of Seveso, criticizes her reliance on stereotype and negative depictions of

Seveso as a community of unsophisticated provincial people, and concludes, "In spite of the publicity launch and the forcibly favorable reviews, the book did not have success . . . a merited failure."[8]

Readers may wonder why I first cite a review from the London-based *Economist* and not another Italian publication. The answer is quite simple: I have been unable to find any Italian-language reviews of *Visto da Seveso* or *Una lepre con la faccia di bambina* from around the time of their publication. Apart from a 1965 review of *La condizione sperimentale*, the extensive Laura Conti archives at the Fondazione Micheletti contain articles about Conti and her work printed only in 1983 or later, and none directly address the two books in question here.

In response to critiques of Conti as an insensitive outsider, I counter that *Visto da Seveso* (written just months after the ICMESA explosion) makes no claims to represent anything other than her own impression of the events as they are unfolding. She writes explicitly from the point of view of someone often present in Seveso but admittedly not a member of the community. The text is not concerned with the actions of "average would-be helpers" so much as with the actions of those who were already involved in the disaster and remediation efforts due to vocation and government appointment, like Conti herself, or another form of local affiliation. Some of her frustration is indeed directed toward the residents of Seveso, but it can always be traced back to the act of communication being hampered by those with the greatest access to information. There is no denying that Conti wrote from an awkward position, as she "saw from Seveso" during the day but was able to return to her home in Milan at night. At the same time, it is hard not to surmise that Conti faced extra critique due to her social position as an unmarried woman scientist, just as Rachel Carson had before her.

Most remarkable to me in *Visto da Seveso* is the text's strikingly forward-looking approach when it comes to the relationship between human and animal bodies and all of the dynamic matter in our midst. Conti explains that she is deeply attentive to the social and environmental position held, in particular, by the members of Seveso's artisanal community. More than the factory worker or the farmer, she writes, it is the local craftsman, the furniture maker, who most clearly articulates the connection between human existence and nonhuman matter in contemporary society. The Seveso artisan labors in his workshop (he is explicitly male in this formulation) in order to craft parts that will then join other parts crafted by other makers, before he or his wife gathers ingredients for a supper from his own backyard. He is as tied to the nonhuman nature surrounding him as he is to the material goods that he fashions with his own hands, and to the neighbors with whom he exchanges those goods.

Ruminating on this figure, Conti writes: "I began to realize that 'environment' is not just the combination of water, air, earth; that one can't consider man in his relationship to nature without also considering his relationship to other men, and his relationship to the objects that he fabricates or the plants that he cultivates."[9] This statement confirms Barca's assessment of Conti's "human centered ecology," while also opening space for it to be reconceived as a human *enmeshed* ecology. By approaching humans as an active, agentive part of the environment, relative to our engagements with other agentive things, Conti moves far beyond staid notions of nature/culture binary. And by considering objects and plants a fundamental part of the dialogue, she also anticipates by decades key concepts in postmillennial environmental humanities scholarship, such as Karen Barad's notion of intra-action, which refers to "the mutual constitution of entangled agencies," and material ecocriticism more broadly.[10]

As Serenella Iovino and Serpil Opperman write: "Developing in bodily forms and in discursive formulations, and arising in coevolutionary landscapes of nature and signs, the stories of matter are everywhere: in the air we breathe, the food we eat, in the things and beings of this world, within and beyond the human realm." Material ecocriticism, they explain, holds that all of the nonhuman world has the capacity for narrative, that "all matter, in other words, is a storied matter.'"[11] As a theoretical approach, material ecocriticism moves beyond an early ecocritical focus on literary text and traditional notions of the natural environment as being outside the realm of communication, in order to examine the stories told by a pile of logs and the ax resting next to them, as well as the fox skulking nearby and the bacteria moving through that fox's belly. By turning our attention to all of this "vibrant matter," to use Jane Bennett's term, material ecocriticism thus removes the human from sole primacy of place, just as it allows us to reinterrogate the relationship between humans and the rest of the multifaceted material world.

As noted above, Conti is keenly attentive to the human. I do not mean to suggest that her work, in *Visto da Seveso* or elsewhere, fully decenters the human in favor of other material agents. Instead, I argue that her focus on our deep enmeshment with everything that surrounds us—and on the capabilities of all those organisms and all that stuff to effect change—looks forward to the material ecocriticism practiced by scholars today. Conti's concern for human and nonhuman animal health, the "love" and "curiosity" referenced above, lends itself to an inclusive and inquisitive gaze upon the lively world at large.

In *Visto da Seveso*, she is particularly attentive to the ability of dioxin to connect to and through both living creatures and nonanimal matter. Shortly after the passage from that text quoted above, she describes a visit with a soil ecologist to heavily exposed territory. He explains that "when the matter that forms the body [of an intoxicated animal] dies, it does not remain still. It will be eaten by other organisms that will in turn die."[12] Describing a simple food chain, the ecologist underscores the fact that the movement of matter and energy from one organism to another can harm as well as nourish. Simultaneously, he alerts Conti, and in turn her readers, that dioxin possesses both great agency and great durability, effecting change as it persists and passes from one vessel to the next. In response, Conti writes, "I saw, in my imagination, the soil as a swarming mass of organisms in movement, of substances in continual transformation."[13] So often overlooked elsewhere, soil is humble but potent stuff in the stories traced by dioxin. It serves as conduit for the toxin to pass through, receptacle for it to linger and thrive, and fertile terrain for regeneration: all processes involving, as Conti notes, transformation.

Transformation takes time, of course, which can be challenging to contemplate in the wake of a crisis. Describing the difficulties she faced in explaining dioxin's durative nature to those affected in Seveso, Conti writes: "Not now, but in three months, maybe a year, an illness will come; in the face of these assertions people laughed at me from their core. The future doesn't exist. An old man, resting his hand on the head of his granddaughter, said to me: 'Dear doctor, even this creature will one day die. But because I don't know *when*, for me it's as though she will *never* die.'"[14] Again, the question arises of how to believe in something for which we do not yet have proof, or how to engage the sort of timescape perspective Barbara Adam proposes, as cited in this work's introduction.

Noting that the idea of "landscape" is inclusive, in that it allows for both a recognition of absences and an understanding of nature-culture overlap, Adam writes: "With the idea of the timescape, I seek to . . . develop an analogous receptiveness to temporal interdependencies and absences, and to grasp environmental phenomena as complex temporal, contextually specific wholes."[15] By focusing on time as a "scape," a representation or angle by which to conceptualize something, Adam asks us to consider socio-environmental phenomena from a perspective beyond the visible, to recognize what is present and what is not, as well as the possibility for eventual transition. It is for this that I find her work so helpful in thinking through environmental change and its impacts on living bodies, and in thinking through processes such as toxicity in particular. It is, in fact, her emphasis on the very notion of *process* that most aids an understanding of dioxin contamination as an event that is ongoing in time, even more than it is moving through space. In *Timescapes of Modernity: The Environment and Invisible Hazards,* Adam cites radiation as just one example of myriad "contemporary phenomena and processes that work invisibly beyond the capacity of our senses." As she explains, radiation "permeates all life forms to varying degrees and disregards conventional boundaries: skin, clothes and walls, cities and nations, the demarcation between the elements," and "its 'materiality' thus falls outside the traditional definition of the real."[16] She could just as easily be referring to dioxin, a substance that, as Laura Conti understands well, is very much outside the realm of the traditional "real" in 1976 Seveso, pushing against boundaries of both materiality and time.

Conti's deep awareness of transformative and temporally durative material flows underscores her other book on the Seveso disaster, *Una lepre con la faccia di bambina*. This short novel is explicitly focused on acts of communication exploring potential futures, and it seeks to access that "emotional connection" that Conti describes in *Visto da Seveso* as difficult to reach. Set in and around Seveso just following the disaster, the book is narrated by a twelve-year-old protagonist, Marco, as he and his friend Sara seek to make sense of their new reality. As she explains in its introduction, Conti had originally planned to write the book for readers around Marco's age. She had hoped to pen an educational tale in the style of *Pinocchio* author Carlo Collodi, designed to offer information about trichlorophenol, dioxin, safety measures, and more. She soon realized, however, that Collodi's model could not apply. He had written in the late 1800s, a time when adults clearly communicated their values to children, often via material indicators of class status, and storybooks served as educational tools. In 1976 Seveso, however, as residents were separated from their homes and communication faltered, traditional channels for the transference of values and education about current events had lost their efficacy.

Conti thus decided to change *Una lepre con la faccia di bambina* from an educational book to "a book about a particular crisis of the educational process," and "from a book for children, it became a book for adults."[17] Her decision reflects a familiar message from *Visto da Seveso,* in which the author notes: "I went about realizing, day by day, the problem of communication."[18] As much as she was concerned with assessing the risks presented by Seveso's dioxin, she was even more alarmed by the challenges it posed to traditional processes of communication within the community. *Una lepre con la faccia di bambina* explores this challenge, as Marco and Sara piece together the tools for comprehending their new reality and potential futures, without the clear assistance of adults.

I am not alone in my interest in *Una lepre con la faccia di bambina*. An eight-page section of the novel translated by Patrick Barron and Anna Re appeared in the journal *ISLE: Interdisciplinary Studies in Literature and Environment* in 2000, as well as in Barron and Re's *Italian Environmental Literature: An Anthology* (2003). Most significantly for my analysis, Serenella Iovino has also written about Conti's work. Through a deft material feminist ecocritical reading, Iovino argues for the "epiphanic" power of dioxin as a narrative agent capable of exposing a vast web of social inequities and material entanglements in both of Conti's Seveso books, but especially the 1978 novel. In her study of Conti's Seveso works, Iovino explores the ways in which dioxin serves as both material and text, "the bodily element that, in its extra-rational materiality, reveals the irrational practices and the cognitive dissonances of an uneven and discriminating society," particularly when it comes to the debates surrounding women's access to abortion.[19] Like the female body itself, she argues, dioxin becomes a site in which the community's struggles to come to terms with environmental toxicity, sexual self-determination, class struggle, and more are revealed. At the same time, the substance demonstrates its own particularly narrative agency, including traits of linearity, chronology, cause and effect.

My analysis builds on Iovino's attention to the revelatory and liberating potential of narrative, by considering instances of narrative practice as modeled within *Una lepre con la faccia di bambina*. The novel's two adolescent protagonists spend much of the text in conversation with one another as they suss out the events unfolding in their community. Their greatest moments of cognition occur through brief unanticipated episodes of role-play and speculative storytelling, what I call "indirect micronarration." I argue that the adolescents' narrative play serves as a model, demonstrating to Conti's adult readers the types of communicative inquiry she found lacking in Seveso's wake: by interjecting her novel with moments of *inner*textual narrative and subsequent understanding, she presents storytelling as a sense-making tool both within and beyond the text.

In this, she again anticipates more recent trends, particularly the turn toward econarratology outlined in this work's introduction. In the words of Markku Lehtimäki, "narratives can open up new, surprising, and thought-provoking insights into the ways in which understandings of the natural world are imbricated with discourse about the environment."[20] Not only do narratives allow us to imagine myriad environments and experiences, they also allow us to examine the ways in which discursive practices surrounding the environment, including human action and nonhuman nature, correspond (or not) to lived experience of the more-than-human world. As Iovino's study demonstrates, Laura Conti's Seveso narratives reveal a disjuncture between residents' lived experience of dioxin following the disaster, which was acute and unnerving, and official discourse in the area surrounding dioxin's toxicity, which was minimizing to say the least.

Erin James argues that an econarratological reading necessitates a merger of contextualist and cognitive narratologies. Whereas the first analyzes narratives within the cultural, historical, and ideological circumstances in which they are produced (Chatman 1990), the second emphasizes readers' intellectual and emotional processing of narratives (Herman 2000, 2002). *Una lepre con la faccia di bambina* lends itself well to both approaches. A fictional narrative based entirely on a real-world occurrence and thus deeply rooted in the Catholic northern Italy of the late 1970s, Conti's text cannot be examined without attention paid to the particular context that inspired it.[21] At the same time, I venture to argue that the work is even more useful to econarratology as a cognitive model, thanks to the use of indirect micronarratives within the text, to be examined further below.

On a cognitive narratological approach, David Herman writes: "Narratives conveyed by oral as well as written traditions can be viewed as a kind of cognition-enhancing logic in their own right, whereby states and events can be arranged into understandable and manipulable patterns; spatiotemporal relations can be established between regions of experience and between objects contained in those regions; relatively distant or intimate perspectives can be adopted; participants can be assigned roles and situated within networks of beliefs, desires, and intentions; and so on."[22] In *Una lepre con la faccia di bambina,* Conti weaves together a chronologically linear story about a community's response to an industrial chemical disaster and the ensuing ambiguous reports of chemical toxicity. With this her text indeed stands as a "kind of cognition-enhancing logic," walking readers through the immediate aftermath of the Seveso disaster and the sense-making process that those affected, in life as well as fiction, had to undergo. So much of that comprehension revolves around the not-quite-perceptible presence of dioxin in their midst. It is this nebulous presence and its relationship to temporality, futurity, and embodiment that steer the content of the narratives *within* Conti's text, and those inner-textual narratives that most directly model the use of narrative practice to arrange patterns, spatiotemporal relations, perspectives, and roles.

Through brief episodes of storytelling, role-play, and speculative projection, the adolescents Marco and Sara acknowledge and seek to understand the presence of dioxin in their midst with a clear-sightedness that others in the book are unable to achieve. *Una lepre con la faccia di bambina* suggests that the hazy materiality of dioxin forces an expanded understanding of both observation and narration. It opens up the first to include the witnessing of secondary effects (if not dioxin floating downstream, then the animals dying from exposure), the overhearing of gossip and half-formed truths, and a careful attention to emotion and affect. It pushes the second to allow for play, brevity, and imagination to take part in the temporal "sequencing of something for somebody" associated with narrative form.[23] In so doing, Conti's novel reaffirms just how important it is to tell the tales of our lived world precisely in its haziest moments.

Marco and Sara turn to imaginative narrative practice as a means to sort out knowledge and perception but also to reaffirm their place in the world. Through episodes of storytelling, speculative projection, and role-play, they shape a personalized understanding of dioxin exposure in their local community and beyond, claiming authorship in an unsettled and increasingly undefined time and place. With very little that has been clearly witnessed to recount, no sensorial obviousness to the heightened presence of dioxin in their environment and bodies, the two adolescents turn instead to the realms of possibility and imagination, actively articulating their own truths as they seek to make sense of changing realities and unknown futures. That their relationship is one founded in narration, in recounting things, is underlined midway through the novel, when Sara says to Marco: "Don't sit there like an idiot. Tell me something (*raccontami qualcosa*)."[24]

As *Una lepre* unfolds, Marco continually tracks Sara down to talk about what is happening in their community. He visits her first at her home in the days following the ICMESA eruption, then later at their shared postevacuation hotel, both before and after she is sent to a hospital to have her chloracne-scarred face bandaged. Throughout their conversations they piece together overheard facts, gossip, and other stories. Early in the novel one of Sara's brothers tells her that all the flies near the factory have been dropping dead from whatever it was that happened that day. Shortly thereafter, Marco hears his mother and one of her well-heeled friends gossip about the "actual" reasons behind the first reported case of chloracne in the area: clearly a low-income family, the women say, scalding its own child in order to profit from rumors of toxicity and receive a factory buyout. A few weeks later Marco listens to a group of older boys explain knowingly that dioxin is what the Americans tossed on someone named "O Ci Min" somewhere far away—a reference, of course, to the revolutionary Vietnamese politician Ho Chi Minh (1890–1969) and U.S. troops' use of the defoliant Agent Orange, a close cousin to dioxin, during the Vietnam War. This final bit of overheard information points to the important relationship between toxicity and transnationality: while the Seveso disaster was played out in ways particular to the local community, especially regarding the question of abortion, concentrated exposure to Persistent Organic Pollutants such as TCDD is squarely a global epidemic.

The friends also tell their own stories, exploring the relationship between individual development and time, and seeking to make sense of the limitless factors that may change the course of a life. They speak about who they were in the past, who they might have otherwise been, and who they might still turn into, always in relationship to the specter of dioxin and always rooted in the physical body—that vessel through which they might eventually perceive further effects of exposure, like the small scars growing on Sara's face. The first node around which Marco and Sara's narrative play centers is based in relationship to animals, the earliest beings to show signs of toxicity after exposure events. The novel opens just as Sara has snuck over to Marco's window one evening to give him her cat Carmelina, who is ill. When Marco asks why, Sara responds, "Oh, right, Saturday you were camping" and proceeds to describe the appearance of a strange cloud outside of the ICMESA factory and the subsequent death of area animals, drawing out cause and effect in narrative form in a way that Marco's mother, for example, refuses to do.[25] From this initial straight story of events, Marco and Sara then move toward tracing out less direct but equally informative brief narratives.

Carmelina deserves special mention here, as do the other animals referenced throughout the book, which is, after all, called "A Hare with the Face of a Child."[26] The title is a reference to the fear that babies born after dioxin exposure in the wake of Seveso might display a cleft lip, a condition that develops during the fetal stage when the tissues of the face don't join properly. In Italian, the common term is *labbro leporino,* corresponding to the once commonly used "harelip" in English. Sara explains to Marco midway through the text that babies born to mothers exposed to dioxin might have the "face of a hare." Again referencing Vietnam and the wartime exposure of residents there to the highly toxic Agent Orange, Marco responds: "Who knows what it must be like to be born in a country where all the children have the face of a hare. Maybe the hares have the face of a child."[27] The exchange takes place during a longer conversation illustrating the friends' differing comprehensions regarding reproduction: savvier to questions of sexual development and biological transfer from mother to child, Sara explains that fetuses "receive blood from their mothers" (as a further indication of both her more advanced development and the anxious conflation of toxicity and female sexuality in Seveso, her chloracne is initially misdiagnosed as hormonally induced acne indicating the impending arrival of her first menses).

Later that night, Marco has a dream involving multiple hybrid creatures: calves mixed with fish, hares mixed with children, and even Sara's cat Carmelina with the face of a hare. This final image—in which Carmelina takes the place of one of the aforementioned children with the face of a hare—marks the climactic use of a trope present throughout the novel: Carmelina as Sara's "little sister," a stand-in for the result of a pregnancy that her mother chose to abort years prior. Such blurring of human-animal boundaries highlights Marco and Sara's expansive views toward animals, which regularly feature as protagonists in their indirect micronarratives. What's more, it demonstrates one of their primary methods of making their way to comprehension about their present reality (here about the potential fate of exposed human fetuses) by playing with related stories and ideas. To use an oft-cited Emily Dickinson line, Marco and Sara regularly work to "tell all the truth but tell it slant"—she perhaps more consciously than he. As the poet warns, "The Truth must dazzle gradually / Or every man be blind."[28] Because the potential for dioxin to mutate bodies in significant ways may be too much to make sense of all at once, Marco and Sara instead dip cautiously into that cognitive and emotional understanding with their own brief experiences of storyworld immersion.

The friends also tell it slant through instances of role-play that can similarly be considered indirect micronarratives. Shortly after the conversation at the book's start, in which Sara explains the ICMESA explosion, they convene again, now to bury a dead rabbit in Sara's backyard garden. After lamenting Carmelina's recent death, as well as what has become a generally dark mood in her home, Sara suddenly turns and says laughingly to Marco: "Would you like a bit of poison, sir? Can I offer you a snack?" She goes on, still in the newly adopted character of amiable host, to explain: "The tomato is a little bit poisoned. I'm sorry, sir, but you know how it is around here. The oil, however, comes from our family land [*casa nostra*]: smell how good it is. The oregano is from our family land, too. My cousin gathers it on a beautifully perfumed hill. When you go up on that hill you smell so strongly of oregano that when you go back home, they mistake you for a tomato salad. Eat, sir, don't be afraid: before I prepared the bread, I washed my hands." Sara then shows Marco her hands, which he observes to be clean except for her black-edged fingernails. In response to that black he says, "I see that you're in mourning, ma'am," to which she replies, "mourning for the dead animals, sir." Marco describes to readers: "She looked at me maliciously, to see if I had the courage to eat the poisoned snack. But I was surely already poisoned . . . and so what did it matter, a bit of poison more or a bit of poison less?"[29]

In this brief moment of somber discursive play, Marco and Sara address toxicity and its normalization, Sara expresses sadness for the loss of beloved animals, and Marco acknowledges for the first time in the novel that TCDD must already be in his body. By slipping seamlessly into role-play Sara recalls Marco Armiero's previously cited claim that "to narrate means to counter-narrate," by verbally skirting around and in fact disproving the dominant local discourse of doubt regarding dioxin and toxicity. She introduces toxicity as fact, rather than dubious hypothesis, into a conversation that would be entirely "real" (as far as readers can tell, she is actually offering Marco the traditional Sicilian snack of tomato on soaked bread) if not for her use of the formal title "sir," which Marco obligingly reciprocates. In her description of the snack Sara gets at the truth of environmental toxicity without overwhelming Marco. She creates a space in which they can experience what it feels like to treat concentrated dioxin exposure and its dangers as an understood reality, without pinning them to a declaration that no one else in their families has yet made. Although brief, her description goes beyond a simple conveyance of information to incorporate linearity, chronology, the suggestion of beginning and conclusion or cause and effect, the completeness imbued by imagination: narrative.

By initiating this moment of dioxin-exploration through imaginative role-play, Sara embodies Martha Nussbaum's work on the relationship between narrative imagination and the development of compassion, which Nussbaum holds essential for participation in community: "Narrative imagination is an essential preparation for moral interaction. Habits of empathy and conjecture conduce to a certain type of citizenship and a certain form of community: one that cultivates a sympathetic responsiveness to another's needs, and understands the way circumstances shape those needs."[30] By imagining themselves as the characters in their play, characters who knowingly inhabit a "poisoned" environment, Marco and Sara develop their awareness and empathy regarding the kinds of decision-making that must occur in such a state. At the same time, they recognize the connection between circumstance and place, while also acknowledging their own real-life susceptibility to dioxin toxicity. Notably, this is the first time in the text in which Marco recognizes that he too may be "poisoned." Such acknowledgment, according to Nussbaum, is also part of compassion, which requires "a sense of one's own vulnerability to misfortune." As she writes, "to respond with compassion, I must be willing to entertain the thought that this suffering person might be me."[31] Modeling Marco's process within the text, Conti suggests a path by which readers may also develop a deeper understanding of dioxin toxicity, as well as compassion toward those affected—including, potentially, ourselves.

In the exchange described above Sara also rhetorically confirms Seveso's dioxin contamination for both Marco and readers in an oblique fashion based in contrast, quickly established via the provenance of the tomato versus that of oil and oregano. In this, hers is a micronarration of opposition. Telling the story of the healthy southern oregano from field to plate, recounting its qualities and journey, she demonstrates what the "lightly poisoned" tomato from Seveso is not. Sara further relies on this same contrast when she draws out a connection between plant health and human health, describing the oregano's perfume as so potent that "when you go back home they mistake you for a tomato salad." Here the qualities of the oregano attach themselves to a human body through mere proximity and touch, a rather pleasant sort of transcorporeality. Although she does not overtly articulate the connection, readers can only imagine the effect that the "poisoned" Seveso tomato might have on a body instead. If one can take on the smell of a plant simply by being near it, just what might be transferred in the process of bodily ingestion? Through both role-play and contrastive narrative, Sara is thus able to establish both the heightened presence and the transmutability of dioxin in Seveso.

As the story progresses, Marco and Sara's attention turns more explicitly to questions of sexual maturity, reproduction, and their very own existence. One day, during a veiled conversation prompted by the local outcry over abortion, an uncomprehending Marco asks Sara just what she is talking about. He explains, "She didn't answer me: she told me a story about cows in a seminary, an old seminary near our town." As Sara tells it, when all the cows on the seminary property were recently pregnant, two gave birth prematurely to stillborn calves. Of those calves, "one had two heads, the other had fish skin" (this tale then makes its way into Marco's aforementioned dream). Here Sara's storytelling is squarely in the realm of fantastic rumor and legend, and yet again she offers a didactic micronarrative, what turns out to be another tale of transference in miniature, in order to help Marco make sense of his newly unfolding reality. To Sara's wild tale he responds, "So even calves, when they're in their mothers' bellies, can take in poison?" Sara confirms by explaining: "They are also animals with breasts. They take the mother's blood before being born—if the mother is poisoned, they get poisoned too, like babies."[32]

Through her narrative description of cause and effect in the mammalian passing on of chemical toxicity, Sara indirectly explains to Marco why there is so much concern about pregnancy in their own human community. Much as in the tomato-oregano conversation, the latter exchange invokes a shared understanding that there are heightened levels of dioxin in their local environment, and that dioxin is toxic. It also allows Marco, as interlocutor, and Sara, as declarative speaker, to take their developing formula one step further: just as harmful properties may pass from plants to the bodies that ingest them, so too may harmful properties pass from a mother's body to an embryo or fetus. This places the friends' attention squarely on female bodies. As Iovino writes, Marco and Sara's conversations reveal "an explicit awareness of the critical role of women in this system of entangled contaminations. Women are affected twice: in their own bodies, and—trans-inter-corporeally—in their babies' bodies."[33] Sara engages indirect micronarration to lead Marco to such awareness, by using cows as her mammalian maternal example. Implicitly grouping human women alongside other "animals with breasts," Sara negates the anthropocentrism otherwise so present in their community, reminding Marco (and readers) that humans are but one species in a vibrant web affected by toxic contamination. At the same time, by not directly naming the women of Seveso, women such as her sister, Sara (and thus Conti's text) performs a protective act, refusing the intense scrutiny otherwise directed toward female bodies in Seveso's ongoing abortion debates.

Returning for a moment to the premise with which I began, I underline that the micronarratives discussed thus far allow Marco and Sara to retain agency over their realities in an uncertain time. Recounting their lived reality, they give shape (fluid) and name ("poison") to something that they cannot *directly* observe but that they know to be present in the current moment and reaching forward into the future. In seeking to make sense of dioxin, Marco and Sara face an uphill battle, not just because they cannot see it with their eyes or smell it with their noses but also because the adults in their community refuse to speak about it. Echoing Conti's reflections in both *Visto da Seveso* and the introduction to *Una lepre con la faccia di bambina,* her fictional narrator Marco explains that the adults in his community "spoke of their businesses that were going bad, of the mayor and the region who had done a lot of stupid things, but no one remembered anything about dead animals and poison. . . . '[G]o, go' they said, 'we don't have anything to say, no one's sick here.'"[34] Such a response is well in keeping with the steps involved in what Javier Auyero and Debora Swistun describe as the "social production of toxic uncertainty": misinformation, shifted responsibility, denial, and "blindness."[35] In the absence of clear information, coupled with a lack of physically observable matter, one response to chemically based environmental disaster is simply to deny it altogether. Another, arguably more productive response is to piece together all that still *can* be known about toxic events, including the existence of feelings like uncertainty and anxiety, by recounting such events to others.

And so it is, through their shared entry into coauthored storyworlds, that the adolescents Marco and Sara are best able to comprehend the events unfolding around and within them. A final textual example, again regarding the potential for newly generated life and continued toxicity, focuses directly on Marco and Sara themselves. Discussing pregnancies Sara's mother had terminated in the past, the two friends conduct the following exchange. They use a simple declarative past tense where other Italian speakers might normally employ a conditional or hypothetical past, and Conti's original text intentionally avoids the use of expected punctuation, such as question marks. The first voice belongs to Marco, who has just learned that Sara's mother had originally wished to have an abortion when pregnant with Sara.

> "Sara."
> "What."
> "If you weren't here what happened."
> "What happened to who. To the dioxin that I got? Someone else got it. . . . What are you scared of, that there wasn't anyone here to get the dioxin? There was, there was. Rest assured that there was."
> "And where was I in this moment, Sara, if you weren't there."
> "Oh, who knows. You were here, or else in another place: outside on your bike, or playing ping pong. You were with a girl who was also called Sara, or something else. A girl who had gotten dioxin, or not. Who knows. And what will you do if I die of leukemia: who knows."[36]

The two friends return to this same speculative conversation and unconventional use of verbal mood days later. When Marco once more marvels at what would have happened had Sara not been born, she laughs and explains: "Here before you there was another girl: a girl small like this," and hunches down with her knees angled out, then says "or else a girl . . . tall like this," and begins walking on her tiptoes.[37]

While these exchanges last no more than a few lines, they too fall into the realm of indirect micronarration. Once again Sara, always a step ahead of Marco, traces out a linear path of cause and effect, confirming that when released into the environment and food chain, toxins impact nearby bodies regardless of their identity. In this she underlines the social leveling power held, within a given range, by materials such as dioxin.[38] What's more, she again presents an alternate reality in miniature, which allows both Marco and her to take stock of their actual conditions. They are there together in that moment; they do exist, despite the fleeting and transmutable nature of their lives.

Marco and Sara place their conversation in the realm of fact, using the simple preterit or imperfect past in Italian rather than a conditional past. Instead of stating that someone else *would have* been affected by "her" dioxin, Sara simply says that someone else *was* affected, presenting the hypothetical alternative as factual past. Conti addresses this linguistic choice in her preface to the novel, explaining: "I attempted to employ a language capable of communicating concrete, essential, not very detailed information; [a language that is] generally not used to specify particularities, and especially not used to describe emotional states."[39] There is no subtlety to Marco and Sara's speech, simply no room for potentials articulated as such through the use of the conditional, which, as Gérard Genette reminds us, "indicates information not confirmed."[40] In avoiding the use of an ambiguous verbal mood, Sara is able to preserve their narrative play as a space in which to overcome uncertainty by employing an authority not found elsewhere.

Describing what might have otherwise been (what "was"), simultaneously playful and somber, Sara suggests both the universality and the individuality of her own physical being: if she were not there with Marco, another girl surely would be, and yet that girl would hold a different form. Through brief sketches of their other potential selves, Sara stacks first the liminality of the embryonic state, then the unpredictability of human life, against the hazily omnipresent toxicity of dioxin. In this she confirms a set of heavy truths: there is no knowing how their bodies and identities will develop, there is no predicting the paths that their lives will take, but there is a certainty to the existence of an imperceptible yet active agent of change in their environment. Of dioxin she says, "rest assured that there was"; of everything else: "who knows."

In her analysis, Iovino stresses that Conti's linguistic choices help to further identify Marco and Sara, already "other" due to their transitional adolescent state, as socially marginal figures and thus linked to the "outsider" that is dioxin: "In this oversimplified linguistic dimension . . . the combination of the dioxin's 'deviant agency' and the children's 'syntactic vision' is the real focal point of the story."[41] This connection between dioxin's "deviance" and the adolescents' particular discursive practices is only strengthened when we consider Conti's text as an ecological model unto itself, in line with Corinne Donly's proposed model of the "econarrative," which reflects both "the present-day reality of environmental crisis," and "new, mutualistic models for approaching nonhumanity."[42] Lehtimäki asks, and James reiterates: "How might an author's concern with a particular kind of ecology motivate the use of specific forms? How can techniques for consciousness presentation, for example, be leveraged to suggest how characters' experiences both shape and are shaped by their engagement with aspects of the natural world?"[43] In response we might say that in the econarrative of *Una lepre con la faccia di bambina*, Conti's attention to dioxin's "deviant" nature is mirrored by Marco and Sara's episodes of indirect micronarration.

Their narrative play inserts itself into the text both suddenly and seamlessly, disruptively and organically, much in the way that dioxin may enter and alter a body. And just as suddenly as they begin, the friends' narratives end, Marco and Sara returning to normal conversation or heading home for supper. Regarding the practice of writing on disaster, Maurice Blanchot calls this sort of fragmentation, "the mark of a coherence all the firmer in that it has to come undone in order to be reached."[44] In their playful and at times wild nature, Marco and Sara's micronarratives allow for a moment of relief from the tension of lived reality, just as they facilitate an already forming coherence of facts and knowledge. At the same time, in their unpredictability and disruption of Conti's macronarrative, they demonstrate yet another way in which dioxin might guide the stories that we tell. In all of this, Conti models two related processes for readers—both the way in which the nonhuman world might shape narrative form and the potential for narrative to lead to greater understanding.

Throughout their brief episodes of indirect micronarration, Conti's adolescent protagonists underline that the world and those of us who inhabit it are in a constant process of existence and emergence. In forcing awareness of bodily process and bodily array, to refer again to Alaimo and Bennett, the very thought of dioxin working its way through our local ecosystems and our own very local bodies engages quite directly with a moment of becoming. This is not the "becoming-imperceptible" of which Deleuze and Guattari write, as helpful as that might sound given the imperceptibility of the substances in question, for that process leads to something akin to death in the philosophers' formulation. Instead, we might consider the process a becoming-material-yet-again, perhaps even a becoming-perceptible-in-new-ways.

Making sense of dioxin and its embodiment encourages an imaginative sort of narrative practice, whether one is the storyteller or her listener; a practice that plays with a constant reevaluation and rewriting of possible truths and is necessarily multiple and inclusive. Dioxin encourages and indeed necessitates narration so that the world all around (and within) us can be better known—even, and especially, as it changes. Simultaneously, dioxin necessitates narration so that narrating subjects and their interlocutors might affirm their own existence through creative engagement with the shifting bounds of that same transmutable world. As Conti's protagonists model, we must pay attention to those narratives and participate in those storyworlds if we are to learn from them.

All that said, it is hard to attain a clear understanding of how *Una lepre con la faccia di bambina* was received at the time of its publication, first by Editori Riuniti in 1978 and then by that same publisher's *Nuova scuola letture* label for students in 1982.[45] As noted previously, the Laura Conti archives at the Fondazione Micheletti contain no print reviews of the novel, and just a few pages of material directly relevant to it: two brief letters from individual readers expressing their appreciation for the text, and a collective letter from the students of class 2G at the Canobbio-Lugano Middle School (1989–90), accompanied by a one-page reading reflection from each student. A comment from one, "this book seemed pretty interesting to me, but sometimes boring," summarizes the students' general sentiment, although I suspect that a group of middle-schoolers might respond similarly to any number of assigned texts.[46]

In much greater abundance at the archives are reviews of the film adaptation of the novel, made by the director Gianni Serra for RAI television in 1988. The film premiered on the state-sponsored Raidue network over the course of two nights, September 22 and 23, at 10:30 p.m. It stars Greco Favel and Barbara Ricci as Marco and Sara, Amanda Sandrelli as Sara's tormented sister, and Franca Rame as Marco's mother. Reviews of Serra's film, to be found at both the Fondazione Micheletti and the extensive Franca Rame and Dario Fo archives, suggest a slightly more enthusiastic reception twelve years after the disaster than that received by Conti's book in its immediate wake.[47] Multiple print reviews even cite the film's "poetic" nature; one states, for example, "but between the threat of dioxin and the reactions, some even racist, of the Seveso residents, Sara and Marco's way of living out the drama emerges almost poetically."[48] Other reviews, such as the following, emphasize the film's political undercurrent, which emerges almost in spite of Serra's direction: "Among the multiple interpretations of the Seveso tragedy, Gianni Serra has privileged the poetic, as though he wanted to avoid the political. But *Una lepre con la faccia di bambina* is a political film, with all that that entails, even if its denunciation of responsible parties is weak and at times evasive."[49]

A small handful of reviews also note that the film's ecological engagements read as particularly timely in September 1988, given the recent outcry over the *Karin B* toxic waste incident just a few weeks prior to the film's premiere.[50] At the risk of diverging from Serra's adaptation, the *Karin B* incident is worth describing here for the resonances it bears to Seveso, and not just as an incident of environmental concern. *Karin B* was the name of a large vessel containing approximately two thousand tons of industrial waste originating primarily in Italy. The toxic matter had been discovered at an illegal dump in Port Koko, Nigeria, then loaded into barrels and onto an Italian-commissioned West German cargo vessel that navigated international waters for weeks in search of a new home. The vessel was finally allowed to dock at an Italian military port and the waste distributed to multiple holding sites in northern Italy, including the city of Modena, for eventual incineration. Coming from myriad industrial sources, the waste contained a wide variety of toxic substances: chief among them were polychlorinated biphenyls (PCBs), a close cousin to dioxins.[51]

Not surprisingly, *Karin B* raised significant alarm worldwide among environmental and social justice advocates, as well as government leaders. A *Washington Times* article from 1988 notes that the incident "underlined the widespread practice of European industrial waste disposal brokers of using Third World nations, where environmental protection regulations are often lax or nonexistent, as dumping grounds." Simultaneously Italian, Nigerian, and transnational, the *Karin B* incident exposed the widespread practice referred to as toxic colonialism, the export of toxic waste to less-developed countries. It also offers yet another example of the wide transitory paths of persistent organic pollutants such as dioxins and PCBs, as well as the ways in which stories from elsewhere (Nigeria, The Sea, some other town) can shed light on events much closer to home, and vice versa. In the wake of *Karin B*, Gianni Serra's Italian television audience was particularly primed to worry about the future effect of toxins seeping through soil and water and into their bodies, and thus to consider Conti's Seveso narrative in a new light. As Franca Rame notes in an interview: "If we consider that [in the time] since Seveso, Chernobyl, Karin B, and the other ships with their poisonous garbage roaming around the Mediterranean are all over the news, *Una lepre con la faccia di bambina* takes on an extremely important significance, both for the events back then and for those that are unfolding today."[52] Rame's emphasis on the text's significance, more so than that of the event that inspired it, is notable. She suggests that serving as receptive audience to Marco and Sara's Seveso story, indeed to their many Seveso stories, is an important process for making sense not only of the ICMESA disaster but also of other more recent disasters involving the spread of persistent organic pollutants such as dioxin.

I have not said much about the actual content of Serra's production, as the script stays very close to Conti's novel. It is for this reason that I offer limited analysis of the film here, focused on one particular element. After its initial four minutes of added explicatory material, Serra's script follows Conti's story to a T, generally using her original dialogues word for word with only the occasional sentence tacked on for further explication.[53] What does stand out as unique to the film, besides Rame's melodramatic performance or the casting of Sara as a physically mature teenager, is a particularly effective use of visual storytelling to represent a series of imagined scenarios. I read these brief episodes as another sort of exploratory narrative in miniature, a sojourn into the realm of "what if" so as to make sense of events unfolding in the actual present.

In an essay on representations of the toxic in twenty-first-century ecodocumentaries, Karl Schoonover explores one of the central questions of this monograph: how to represent nonvisible toxicity. He notes that "the visibility of toxicity is tricky because it permeates our world in ways that are hard to be shown in the image" and emphasizes the limits of photographic mediums for their inability to "show us toxicity in action."[54] Pointedly, he concludes that "the toxic challenges cinema's capacity to render the material world as visual text, and in doing so asks whether cinema is still the medium to redeem the physical world in our eyes."[55] While Serra's production is far from an ecodocumentary, it is a form of ecocinema—in this case televised—and it does make a case for the ways in which visual texts might continue to "render the material world as visual text," precisely by engaging in a uniquely visual form of indirect micronarrative. Three times throughout the film, Serra veers from a relatively realist mode into brief unannounced fantasy sequences reflecting Marco's internal visions. In this, Serra's *Una lepre* furthers the work of Conti's not only by bringing it to a presumably much larger audience, Italy's television viewing public, but also by adding elements unique to the visual medium, thus making for a different and equally potent form of storyworld immersion.

The transitions into these sequences are seamless: as Marco moves through the day, he pauses midaction while an imagined scenario begins to play out before him, with no formal indications of the change in register other than his widened eyes. The first such vision occurs just ten minutes into the film. It is the morning after Sara has told Marco of the explosion and brought him her cat Carmelina. With the sick animal tucked into his jacket, Marco gets on his bike to ride over to Sara's house. As he approaches her factory-proximate neighborhood, he stops at a red light and, looking ahead, sees the street littered with dead human bodies, bags of groceries sprawled out next to them. The camera shifts focus and pans low to reveal the street before him, moving across dozens of people of all ages who lie motionless, as though they had all collapsed in unison. The camera then pauses at the sight of Sara, sitting upright on her knees as she cradles the head of a fallen boy who, as focus sharpens, appears to be Marco himself. As she lifts her gaze to the camera, it cuts back to frame the actual Marco still waiting at the stoplight on his bike. He too lifts his gaze just as the light turns green, and peddles on to continue with his actual waking day, now passing in front of the camera, which pans to the left to follow him from behind, revealing a perfectly empty street, not another person in sight.

This brief scene, like those revealing his other visions in the film (first of Sara's sister in flowing gown with head wrapped like a beatific mummy, then of a dead Sara, tossed like garbage by sanitation workers at a cleanup site), wordlessly conveys the sense-making happening in Marco's brain. It comes the morning after Marco has learned of the explosion and the death of animals in Sara's neighborhood, to which he is traveling for the first time since the event. Like the film's establishing scene or occasional additional dialogue, both added to convey further information to viewers, Marco's visions read as a none-too-subtle aid for those of us at home, letting us know that he is afraid and worried about what might happen next.

Entirely lacking in dialogue or voice-over narration, such scenes rely almost solely on the visual, backed by a tense instrumental score. This reliance on visual and auditory perception emphasizes the sensorial capacities of film or television, thanks in the aforementioned scene to point-of-view editing, which prompts a viewer's gaze to be aligned with Marco's. Elena Past has argued, also in the context of ecodocumentary, that sensorial connection with on-screen subjects can help to communicate the affective experience of toxicity. She cites Vivian Sobchack's work on the "cinesthetic subject" who has "a bodily experience of the senses while watching a visual text." As Sobchack writes: "Because our consciousness is not directed toward our own bodies but toward the film's world, we are caught up without a thought (because our thoughts are 'elsewhere') in this vacillating and reversible sensual structure that both differentiates and connects the sense of my literal body to the sense of the figurative bodies and objects I see on the screen."⁵⁶ Watching an on-screen subject walk through a contaminated field and scrunch up her nose, for example, we might imagine what it is to see that landscape up close, to smell whatever odor is present, and to contemplate the imperceptible changes that the contaminating substances might enact on creaturely bodies. In the aforementioned scene from Serra's *Una lepre,* our field of vision is aligned with Marco's such that we see what he sees not just before him but also in his head, his own personal indirect and exploratory micronarration that necessarily pushes the limits of just what we might consider to be the "physical world," to refer back to Schoonover's critique. Extending Past and Sobchack's arguments on the sensorial, I argue that this brief scene allows us to share in Marco's experience of imagining, and thus his expanded understanding of materiality.

In an econarratological context, and in the spirit of indirect micronarratives, such visions model the interior workings of Marco's cognitive and emotional response to the disaster in his community, as he pieces together the bits of information he has learned, simultaneously exploring their potential outcomes and testing the limits of his fears. Like the brief stories he and Sara share in Conti's original text, Marco's visions also allow receptive audience members to critically assess how likely his feared outcomes might really be, and to imagine what our own fears might be in such a scenario. Both steps ask us to draw upon whatever knowledge we may have about dioxin, thus engaging in our own processes of sense-making just as we take part in, to cite James, "the mental modeling and emotional inhabitation that underlie narrative comprehension."⁵⁷

Una lepre con la faccia di bambina, whether as book or televisual adaptation, makes deft use of story and imagination within the text, so as to show the benefits of the narrative process for those, like us, who exist beyond it. As it does so, it conveys information about a significant environmental disaster, the effects of which are still being assessed to this day. *Una lepre* is, of course, not the only narrative text about the Seveso disaster. In addition to the scholarly, journalistic, or memoir-based works, such as Rocca's, already addressed at the start of this chapter, I would be remiss not to mention two more recent international cinematic treatments of Seveso. Sabine Gisinger's documentary *Gambit* (Germany, 2005) tells the story of Jorg Sambeth, ICMESA's technical director at the time of the disaster, and his personal investigation into the corporation's behavior in its wake. It is one of few texts I have encountered that is focused primarily on ICMESA. It raises interesting questions about the difficult overlaps between corporate and personal responsibility, while also addressing the disaster's transnational nature, as well as lingering Cold War–era conspiracy theories about ICMESA's possible production of agents for chemical warfare.

More intriguing from an econarratological perspective is the 2001 Japanese animation film *Inochi no Chikyuu: Dioxin no natsu* (translated online as both "The life on Earth—*The summer of dioxin*" and "Tracing the gray summer") by Satoshi Dezaki. This heavily fictionalized film bears certain parallels to Conti's story in that it features a group of young friends in Seveso as they seek to make sense of the disaster, the new environmental risks it poses, and questions of reproductive futurity. Where this film greatly diverges from Conti's text is in its focus on the adolescent protagonists' active sleuthing, as well as its final resolution, which sees them publicly shaming Hoffman La Roche executives in a moment of great triumph—a redemptive rewriting of the original narrative. Special mention ought also to be made of Umberto Lenzi's 1980 film *Nightmare City*, which is addressed further in chapter 5. Like similar films produced at the start of the 1980s, *Nightmare City* speaks to Cold War–era concerns about nuclear contamination and the end of civil society, while utilizing all the campy tropes of a cheaply made zombie slasher set in an ambiguously Western city. At the same time, it draws from the recent Seveso disaster, most overtly through representations of extreme chloracne, to both play on audience anxiety and deliver a clunky environmentalist message.

There is certainly more to say about these films, as well as the various television specials that aired in the years following the Seveso disaster, whether in Italy or elsewhere. I privilege Conti's work, however, for multiple reasons. Although relatively unknown to English-speaking audiences, she is one of Italy's preeminent twentieth-century environmentalists, and her ecologically engaged writing looks forward to twenty-first-century critical approaches attuned to the deep enmeshment between human bodies and the more-than-human world. Furthermore, *Visto da Seveso* is the most detailed description of the disaster and its aftermath in print, and universally serves as the primary reference for more recent studies, such as those by Centemeri and Fratter. Most significantly, however, Conti's attention to the power of storyworld immersion in *Una lepre con la faccia di bambina*—her attention to the possibility for narrative to lead to knowledge—sets the framework for what emerges decades later in the stories being told about the crisis still unfolding in Taranto to this very day.

FIRST PERSON

Massimiliano Fratter

From a 2015 conversation with Massimiliano (Max) Fratter, who at the time directed programs at Seveso's Oak Forest. We had spent the afternoon together walking the park grounds, one day before my meeting with the Circolo Legambiente Laura Conti Seveso, in which he also participated.

> MAX: Everyone has their own way of telling the story. It's unique also because we're here, physically, we were born here. . . . History cannot be recounted in a solely objective way, you know, I told you this before, right?
>
> You can't think that this place wasn't also—and perhaps for some people still is—a place of pain. Of struggle, sure, but also of pain.

May 19, 2015

CHAPTER 3

Taranto
Past, Present, Future

Taranto, as anyone will tell you, is a place of great contradiction. It is home to deeply rooted artisanal and maritime traditions, stunning natural beauty, and strong community ties. For more than half a century, it has also been home to shockingly high levels of industrial pollution, palpable anxiety regarding employment and economy, and widespread human illness. If you spend some time in the waterfront city, you will find that many residents are frustrated or angry, while others are resigned to the status quo. Keep exploring, however, and you will soon discover that a great many Tarantini are also incredibly resilient, pushing back against the damage continually inflicted upon their land, bodies, and pride of place. They do so through community networks, social action, and an expanding corpus of creative texts with a clear emphasis on multiplicity of experience and collectivity of voice. Very much in the footsteps of Laura Conti, Taranto's many present-day narrators use story as a means of comprehension and communication. They also use it as a form of protest and survival, for Taranto's is a crisis still actively underway.

Taranto has always been tied to story, boasting two epically entwined foundational myths. As the first goes, the land forming the modern-day city is named for Taras, son of the god Poseidon and the nymph Satyrion. The young Taras was rescued at sea from a sinking ship, thanks to a dolphin sent by his father. He rode this dolphin across the Mediterranean until arriving at a particularly beautiful stretch of curving coastline, where he founded a Greek colony in his own name. The land is said to have then been rediscovered centuries later (circa 700 BC) by a group of Spartan outcasts known as Parteni and guided by the brave warrior Falanto in their search for a new home. After first seeking counsel from the Oracle of Delphi,

Falanto led his companions out to sea for what soon became a stormy journey. Their ship crashed, and all aboard almost perished, but Falanto was swiftly lifted out of the choppy waters by a dolphin (yes, another dolphin) and subsequently able to rescue his fellow voyagers. They eventually found land just where Taras had before them, deciding to stay put and name their new home "Taranto" in his honor.

In the various extended versions one can find of these connected stories, Taras and Falanto both read as captivating figures, to be sure. But the real heroes are the dolphins who brought the men to safety out of a sea that would have otherwise led to death, thus enabling them to find fertile land and nurture a productive home for themselves and others. The coordinates at play in each of these dolphin encounters—passage between realms of life and death, connection to the more-than-human world, access to fertile land and renewed civilization—are all in keeping with the symbolism regularly attached to dolphins in various cultural traditions. The frequent appearance of the creatures in ancient art and myth presents them as a positive omen particularly tied to seafaring, as well as a point of connection to life, death, and rebirth. In Phoenician tradition dolphins often appear alongside Atargatis, the mother goddess of vegetation, sentient life, and rebirth. Similarly, Greek mythology associates dolphins with Dionysus, who every year dies only to be reborn once more as the god of vegetation and fertility. Dolphins also feature as friends and saviors to many other young male protagonists in Greek lore, from Arion to Theseus. The creatures make their way into Christian tradition as well, with no fewer than five different saints rescued at sea by dolphins, who are variously interpreted in the associated literature as Christological symbols, emblems of rebirth, or auguries of love.

Thus, as in so many tales and traditions, the strong associations between dolphins and Taranto's origins speak to a vibrant realm of ebb and flow, natural (re)generation, and connection to the more-than-human, in the form of other living creatures but also via the afterworld. These coordinates help set the tone for Taranto's history in subsequent centuries, as a capital of both maritime and land-based trade and cultivation (a symbiosis well evidenced by the archaeological remains stored at MarTA, the National Archaeological Museum of Taranto). They also set the tone for Taranto's more recent decades, including its present-day realities, as a land and community intimately aware of the connection between all sorts of creatures and material substances, as well as the ever-present possibility of death.

As noted in this book's introduction, Taranto has developed over the past sixty years as one of contemporary Italy's largest industrial centers, due to its strategic coastal setting along the Ionian portion of the Mediterranean Sea, encircling an internal bay known as the Mar Piccolo. In turn, it has also become one of Italy's most significant sites of toxic emissions.

Although Taranto has been home to a major naval base since the mid-1800s, the city's industrial culture and identity were truly shaped toward the end of Italy's post–World War II economic boom period. The first half of the 1960s saw the development of three large enterprises that would bring significant shifts to Taranto's population and economy. In 1964 the government holding company Industrial Reconstruction Institute (IRI) built a sizable cement plant known as Cementir (later sold to the Caltagirone group in 1992, then again to Italcementi in 2018). That same year, 1964, also saw the development of the region's largest oil refinery, first run by the multinational Shell group and then eventually taken over by Italy's Eni (formerly Agip) in 1974. One year later, in 1965, the Italsider steelworks became operational following a massive construction plan first launched six years prior. Like Cementir, Italsider was initially supported by the state-sponsored IRI. The cement plant, oil refinery, and naval base were, and indeed still are, impactful in their own right regarding labor, economy, and environment. This is especially true regarding the Eni refinery, which has recently begun processing crude oil transported from the Tempa Rossa oil fields in the Basilicata region, thanks to a large pipeline completed in 2019.[1]

It was the steelworks, however, to soon dominate Taranto's industrial activity, employ the largest percentage of its inhabitants, and most significantly reshape its residential and agricultural geographies. In his thorough history of Italsider, Tonio Attino writes that 15,000 workers helped to build the initial structures, which stretched out over 600 hectares, or the space required for 840 soccer fields. A substantial government-funded project in the early 1970s then extended the steel manufacturer's territory to 1,500 hectares, or approximately 2,100 soccer fields.[2] Taranto's population grew in kind, reaching a peak of 260,000 people in the late 1970s, up from 150,000–160,000 people in the mid-1950s.[3] This growth was accompanied by a largely unregulated building boom, as additional infrastructure was quickly constructed quite near factory grounds, as well as further out into the diminishing countryside.[4] In 1972 the environmentalist and cultural critic Antonio Cederna declared Taranto, now dominated by constant building and production, "a destroyed city, a Manhattan of underdevelopment and abusive construction."⁵ This was only thirteen years after another noted cultural critic, the writer and filmmaker Pier Paolo Pasolini, had written of what he called "Taranto, perfect city" as "shining between the two seas like a giant smashed diamond."⁶ While perhaps not as effusive as Pasolini, many residents were still more optimistic than Cederna regarding Italsider and its effects. They were heartened by surging levels of employment and economic growth throughout the region, as well as the various cultural initiatives, such as clubs for soccer and singing, launched by the steelworks and common to factory culture at the time.⁷ In spite of various shifts in the international steel market throughout the 1970s, Italsider remained strong, directly employing 22,000 people in 1980, with another 15,000 in satellite roles.

The following decade was more challenging, and in 1989 the steelworks went through the first of a series of changes designed to address what had become a worsening economic return. Under the continued direction of the IRI, Italsider regrouped and took on a new name: Ilva, the original Latin name for the Tuscan island Elba, which was once heralded for its natural iron resources. Six years later, in 1995, the private Riva Group, run by the family of the same name, stepped in to purchase the steelworks, bringing it back to profitability and an annual output of approximately eight million tons of steel. The Riva purchase also included other Ilva sites, including a smaller steelworks far up the coast in Genoa.

In late 2018, Ilva once again entered into new ownership, after six years of heated debate and significant legal and financial woes related to nearly six decades of gross environmental damages. It all erupted in July 2012, following three years of investigation into the health effects of the plant's emissions on the local community, when a Taranto-based judge ordered Ilva's most-polluting blast furnaces to be temporarily shut down. The same judge also placed eight Ilva managers, including members of the Riva family, under house arrest for having knowingly created an environmental disaster over the previous seventeen years. The Italian Council of Ministers had already declared in 1997 that the steelworks represented an "area at high risk of environmental crisis," but, as of 2012, Ilva management had not taken adequate steps to address that risk.⁸

Following the 2012 verdict, Ilva presented a plan for improved environmental measures, but this, too, was declared to be inadequate. In November of that year the state seized the steelworks, threatening it with permanent closure. Significant protest ensued, most notably from workers themselves, who argued that permanent closure of Ilva-Taranto would destroy the local economy. During these months Taranto received considerable coverage by the international press and the primary issue emerged in both the local community and far-flung media as one of jobs versus health.[9] The steelworks remained under government conservatorship, becoming fully administered by the state in 2015, but was never actually shut down, despite the occasional partial closing as officials tried to address continued high toxicity in daily emissions.

In 2017, the multinational steel manufacturer ArcelorMittal won a bid to buy out all of Ilva's production sites at €1.8 billion. The purchase was protracted for a year, and provisionally brought to conclusion under the direction of Deputy Prime Minister and Minister of Finance Luigi Di Maio on September 6, 2018, when ArcelorMittal signed an agreement with trade unions that committed to the eventual rehire of every one of Ilva's extant employees. Local advocacy groups immediately raised concerns that the agreement required no significant change regarding the steelwork's environmental or safety practices, but the global giant's ownership of Ilva officially went into effect on November 1. As ArcelorMittal's negotiations with union and government officials unfolded throughout 2018, members of the Riva family and other former Ilva executives remained on trial for deliberately causing environmental disaster. In early March 2019, ARPA Puglia, a regional environmental agency, released data showing that from 2017 to 2018 the amount of dioxin in Taranto had increased by 916 percent, nearly reaching the historic highs of ten years prior.[10] A few days later, a group of citizens symbolically "closed" the steelworks, lacing a padlocked chain through its main entrance gates.

In the summer of 2018, I had the opportunity to attend one day of the Ilva court proceedings. Referred to as Ambiente Svenduto (Sold-Out Environment), the proceedings were held in a stark cement courthouse in Taranto's Paolo Sesto neighborhood. I observed from the courtroom's almost empty public seating area next to Angelo Fornaro, the soft-spoken patriarch of the Fornaro family, whose ranch borders Ilva grounds. In 2008, the Fornaros' six hundred–plus head of sheep were slaughtered after testing revealed dangerously high levels of dioxin in the animals' bodies. On the day that I attended the trial, the opening roll call of parties involved, including the Fornaro family, took over an hour, but little else was accomplished due to a constant voicing of objections from all sides. As of this writing, the case is still ongoing.

If you visit Taranto, you can easily spot the Ilva plant extending for miles immediately beyond the densely populated Tamburi neighborhood, right behind the central train station. Its towers can be seen from just about any vantage point in the city, and its red steel particles notoriously paint the sky at sunset and leave a fine trace of visible dust across the stationary surfaces of everyday life: parked cars, patios, and playgrounds.[11] Ilva-Taranto is still the largest steelworks in Italy and, by many accounts, in the EU at large, producing over 30 percent of Italy's annual steel output. In 2016 it accounted for approximately 75 percent of Taranto's GDP.[12] Over the course of the past decade, the steelworks has directly employed around twelve thousand people annually, with another ten thousand or so employed in auxiliary roles.

Alongside steel, it still produces toxic pollutants on a daily basis, and residents of Taranto continue to suffer from increased risk of mortality due to cancer and other illnesses, including respiratory, cardiovascular, and digestive disease. The SENTIERI project (National Epidemiological Study of Territories and Settlements Exposed to Risks from Pollution) has been monitoring mortality and illness in Taranto for almost thirty years now. The first phase of the Taranto study was conducted from 1995 to 2002. It showed that, when compared to the national average, Taranto residents exhibited an excess of 15 percent in general cancer incidence, and an excess of 30 percent in lung cancer mortality. Since that time, subsequent phases have shown slight improvements in those numbers, but the rates of serious illness involving lungs, bladder, liver, pancreas, blood, and kidneys remain disproportionally high for adults in Taranto and surrounding areas, and childhood leukemia seems to be on the rise. The fifth phase of the SENTIERI study reported six hundred babies born with physical malformations in Taranto between the years 2002 and 2015, as well as more than forty infants one year or younger with some sort of tumor.[13]

As Stefania Barca and Emanuele Leonardi write, in the terminology of environmental justice Taranto can be considered a "sacrifice zone" of industrial development, its population a "discriminated community, whose right to a safe and clean environment has been disregarded and heavily discounted in politico-economic terms."[14] A decades-long case of environmental injustice has unfolded in the community, as Taranto's residents have suffered from toxic pollution against the threat that the pollution's only remedy would be total Ilva shutdown and, thus, mass unemployment. Both the greatest impact to health and the greatest fear regarding potential job loss have been felt by those who work at Ilva or live in the working-class Tamburi and Paolo Sesto neighborhoods closest to the steelworks.

It is this double-edged suffering that led many, including Ilva workers, to protest against a possible Ilva closure in 2012, while a great many others, again including Ilva workers, advocated for it. That season of protest revealed a deep and still extant divide within the community regarding the feasibility of a Taranto without Ilva and launched important conversations about possible futures and alternative forms of employment. Although public protest and community action may have been at its most vocal peak in the latter part of that year, it continues into the present day, often through organized groups such as the Comitato Cittadini e Lavoratori Liberi e Pensanti (Committee of Free and Thinking Citizens and Workers) and PeaceLink, a volunteer network dedicated to various causes including disarmament, immigrant welfare, cyber culture, and environmental health.

The environmental health risks that residents face are the result of much more than just the dioxin on which this book focuses. Alongside over 80–90 percent of all the dioxins produced in Italy, Ilva's daily emissions also include heavy metal particles, polychlorinated biphenyls (PCBs), sulfuric acid, and more. An air-quality survey published as part of the 2012 court proceedings revealed that, in 2010, Ilva had released more than 4,000 tons of mineral dust, 11,000 tons of nitrogen dioxide, 11,300 tons of sulfur dioxide, 7,000 tons of hydrochloric acid, 1.3 tons of benzene, 150 kilograms of Polycyclic Aromatic Hydrocarbons, and 52.5 grams of benzo(a)pyrene into the air. And 2010 was also the year in which municipal authorities officially prohibited children from playing outdoors in Tamburi, due to the high level of toxic matter found in the neighborhood's soil and on the surfaces of its various structures.

While the Ilva steelworks has been emitting the above-listed slew of substances into the local environment for decades, dioxin is regularly cited by local watch groups and both national and international media sources as the most troubling of the bunch. This is due to its known ability to cause cancers and other grave health problems, as well as the incredibly high levels of the pollutant detected in Taranto's human and animal bodies in recent years. A 2014 microstudy by the environmental group Fondo Antidiossina (Anti-Dioxin Fund), for example, revealed the presence of dioxins and dioxin-like compounds in Tarantine women's breast milk at 1,500 percent of permitted levels.[15] In 2008, state-ordered analysis of the aforementioned Fornaro family's sheep, along with animals from several other properties located within a ten-kilometer radius of Ilva, led to a total of 1,200 animals being sent to slaughter and disposal under the label *rifiuto di tipo 1*. As Giuliano Foschini writes, such a label indicates "highly dangerous toxicity," a confirmation of the animals' status as *"le italianissime bestie alla diossina,"* typical of the region (poorly translated: the oh-so-Italian animals à la dioxin).[16]

Dioxin also, of course, creates an overt connection between Seveso and Taranto. As noted in this work's introduction, the two communities are far from one another geographically, and the temporal realities of their disasters are obviously quite disparate: the ICMESA explosion at Seveso represents an isolated moment of intense dioxin exposure, whereas Taranto has suffered from a quotidian slow violence, bearing the brunt of Ilva's dioxin-rich emissions for decades. In spite of this significant difference, the long-term effects of dioxin exposure are yet to be fully known for either community, given the substance's significant half-life and, even more notably, its ability to pass down to subsequent generations.

In the words of the Tarantine pharmacist, activist, and writer Daniela Spera: "They [dioxins] don't go away! Far from it! In fact, they get worse. And the continued low dose exposure, that's what eventually causes genitotoxic harm—what causes genetic mutations and increased vulnerability to specific pathologies that lead to children being born already with tumors."[17] Spera explains that, while important, the monitoring of spikes in harmful daily emissions from Ilva's chimneys (or so-called "wind days") performed by some local environmental advocacy groups could also be misleading. By focusing attention on immediate experience, "wind days" obfuscate dioxins' cumulative and temporally durative harm, such that the day following spikes, "people say, 'Good! Today you can feel safe because levels are calm, nothing to worry about.'" She reiterates: "We live in a state of daily health crisis—every single day—and you can't just tell people, 'oh, today close your windows!' . . . No one says, 'Watch out because those dioxins emitted into the atmosphere today, and the ones that will get added tomorrow, they don't disappear'—in the sense that they accumulate, day after day, day after day. It's abhorrent."[18] Alessandro Marescotti, the director of PeaceLink, echoes Spera's warning. Although he actively supports the practice of alerting residents to particularly high "wind days," he also notes that "dioxin enters the system and either it never leaves or it takes many years to work its way through."[19] Marescotti has stated elsewhere that Taranto's annual dioxin output is equal to that of Spain, the UK, Austria, and Switzerland combined.[20]

Both Spera and Marescotti speak of dioxin's strong perseverance and temporally durative nature, alongside its ability to travel through *emissioni diffuse e fuggitive,* widespread runaway emissions. Such conversations bring us back to the fact that the process of corporeal transference between generations of humans or animals is deeply tied to the corporeal transference between, indeed the deep enmeshment of, humans and other animals with so much of the material word. Here again it is helpful to consider Alaimo's notion of transcorporeality, which she has most recently defined as a "posthumanist mode of new materialism and material feminism," recognizing that "all creatures, as embodied beings, are intermeshed with the dynamic, material world, which crosses through them, transforms them, and is transformed by them."[21] Such a mode helps to avoid the short-term thinking of which Spera warns, as it necessarily brings subjects beyond immediate place and time and into a much vaster web of relationality.

Alaimo links her reading of the transcorporeal subject, "generated through and entangled with biological, technological, economic, social, political and other systems, processes, and events," to Rosi Braidotti's notion of the transversal subject, which is also informed by feminist principles of situatedness and horizontality.[22] Braidotti operates from a vital, neomaterialist stance, one that acknowledges affirmative flows across boundaries between things and beings. She writes that thinking in terms of "transversal subjectivities" allows us to "think across previously segregated species, categories and domains" along a complex nature-culture continuum of emergence.[23] Much like Alaimo's transcorporeal ethics, a transversal approach decenters the human and allows us to conceptualize subjectivity, or being, as an assemblage based on an "immanence of relations," to again cite Braidotti.[24]

I do not wish to suggest here that Taranto's residents are on their way to becoming Victor Frankenstein's Creature—simultaneously human and other, grossly yet gently beyond the norm. Rather, I wish to emphasize that to be human is to also *be* other, ever in transition to something new. As Eva Hayward writes, that shared prefix "trans" is "weighted with across, beyond, through (into another state or place—*elsewhere*)."[25] We all inhabit bodies that are open to the flows of the substances we encounter, incorporating and shifting with them, whether they help us to grow stronger or to suffer, as our future selves continue to emerge. As such, an attention to dioxin offers a means of approach to the difficult realities of contemporary Taranto, while also serving as a connective thread between the Seveso disaster, the ongoing eco-corporeal crisis in Taranto, and the experience of being in the world at large.

When we recognize dioxin's ability to move through the lived environment and into our bodies—as well as any other bodies we may yet bear—we inevitably confront the deep permeability between living beings and the various forms of matter with which we come into contact all day long. In this light, the story of Taranto's dioxin-rich transcorporeality becomes yet another tale of border crossing in a contemporary world so marked by the presence of imagined boundaries, whether they are attached to an idea of the nation or to that of a body as closed entity. At the same time, it emphasizes the ways in which human subjectivity is inevitably linked to the dynamic and agentive more-than-human world, and welcomes a feminist materialistic critique recognizing the intra-actions between meaning and matter, to refer again to Karen Barad.

As noted in this work's introduction, thinking in terms of transcorporeality and transversal subjectivity can provoke a receptive experience particularly attuned to the ways in which the nonhuman or more-than-human world might also affect a narrative text, shaping not only its content but also its very structure. It allows us to consider the ways in which the patterns and temporalities of the more-than-human world—whether in the form of what we have traditionally considered "nature," such as passing clouds or a flowing river, or something more overtly tied to human manufacturing, such as dioxin—might shape the body of a text: its patterns and temporalities, the length of its sentences, the deep or shallow focus of its lens. Just as *we* are enmeshed with and shaped by the vibrant matter all around us, so too are the stories that we produce and consume, in ways more and less overt.

FIRST PERSON

Daniela Spera

I first visited Taranto in the summer of 2016. One of my goals for that trip, in addition to simply seeing Taranto firsthand, was speaking with the pharmacist, activist, and writer Daniela Spera. Spera has been instrumental in directing international attention to the crisis in Taranto, most notably through her work to bring the community's first collective appeal against Ilva to the European Union Court of Human Rights in 2013. She is also, as discussed further in chapter 4, the inspiration for Cristina Zagaria's fact-based novel "Veleno" (2013, Poison). Daniela and I met in the afternoon at an outdoor café on the edge of Taranto's stately Piazza Immacolata. We fell into a long conversation that took us through a late dinner of street-side panzerotti, hand-held pockets of fried dough stuffed with mozzarella. What follows is a condensed selection of Daniela's comments.

> DANIELA: We have accomplished so many things here in Taranto, and so I feel like I have done what I wanted to do: I have conducted research in the field and provided information in terms of pollution, especially dioxin. Because dioxin and all the other persistent pollutants have characteristics that other pollutants don't have. One can have the sense, analyzing the blast furnaces, the legal limits, going to take deposimetric measurements, and so on, that everything's okay. But no one goes around telling people, "Look, the dioxins that today are emitted into the atmosphere, and those from tomorrow that get added to those from today, they don't disappear." In the sense that they accumulate, day after day, day after day. And so it matters little if an entity has respected the legal limits, because those dioxins are accumulating, and it's that accumulation that puts everything in

motion. For this reason, I've tried to explain to Tarantini that we're being deceived. We don't recognize this [chemical] presence, and the powers that be tell us it's okay. They reassure us, but they're actually just making fun of us, especially regarding dioxins. Because dioxins attach themselves to microparticles, which are what we then breathe, what we find on the land, *in* the land, what makes its way into the water-bearing stratum of the earth.

It all has to do with the fact that the legal limits for pollutants are established in agreement with big industry, that the environmental laws are in agreement with big industry. Otherwise they wouldn't be able to produce, because the production of these substances, the presence of these industries [big enough to have] an environmental impact, practically necessitates pollution. And so there you are polluting in agreement with the law, and it goes unspoken that you are nevertheless causing negative health impacts. It's not the case that we simply need to apply environmental regulations, because there's no connection between a respect for the legal limits and protection of public health—they are two things that don't travel in a parallel fashion; they're divergent. That is to say it matters little, because if you want to really evaluate how much an entity is causing harm, you have to follow the epidemiology, not just cancers, but all of the pathologies. Because dioxin and so many other pollutants cause, in particular, autoimmune diseases—and there are tons here in Taranto: people who have multiple sclerosis, Alzheimer or Parkinson disease at a young age (due to heavy metals), various correlated problems, rheumatoid arthritis. Very young people.

Images are one thing that really captures the public opinion. I saw something that really struck me: it was a program by some national TV station that showed the prevention department from ASL Taranto going to this farm and taking all of the sheep. *All* of the sheep from this farm, and they took them to the butcher. But what struck me the most was the desperation of the farmers. They said: "Is this how you want to resolve Taranto's problems? Killing sheep, goats, letting those who caused all of this carry on?" At that time I had not yet really gotten involved, and I said: "But how is this possible? I am watching this video online yet in Taranto people are calm."

In Italy no one talks about it. There's just this TV program in which they take away six hundred sheep, bringing them to slaughter because they're contaminated by dioxin, and everyone's calm.

[In the rest of Italy] people see the question, as it is presented to them on TV, as a question of jobs. They don't know what the reality really is in Taranto. If you go to the Tamburi neighborhood, every family there has terrible stories to share.

Two or three years ago I brought a scholar from Lecce on a guided tour through Tamburi and we went inside a bar. At a certain point the barista began to share.

"Will you tell us a story?"

"Oh, here in Tamburi, it's a dark situation, each of us has at least one person at home who is seriously ill. The other day one of my neighbors was talking about a child who had died of cancer and only had his torso left: because of the cancer they'd amputated his arms, amputated his legs . . ."

I mean this is how we came to understand what it was like. She was telling this story as though it were just normal. As though it were, I don't know, a flu, no big deal, a fever, totally normal. Just his torso, his *torso,* and he died like this, a child! Perhaps it's better for those of us in this fight to not know all the details. Because otherwise these things block you, they can block the struggle because . . . you cannot be unaffected. You fight, you forge ahead . . . it's enough to know that this situation exists, for you to fight with determination. I think that if we had to really go to the depths of each family's drama, it would just be too much because then the shock takes over.

In the end, dioxin is almost "helpful" because it has digital imprints. Every industrial process produces dioxin imprints: we talk about "dioxin profiles." So, an incinerator produces dioxins, but the imprint, or rather the various congeners that it contains, are different than what you find if you conduct a profile analysis of dioxins from, say, a steelwork's furnace. Just like the dioxin profile from the furnace, from the dioxins that escape from a steelwork's furnace, is different from the profile contained in the electro-filters used to block the dioxins before they escape out into the atmosphere.

How was it possible in Taranto to identify Ilva [as the source of the dioxins]? Thanks precisely to these imprints that are unique to dioxins. PCBs also have an imprint, but heavy metals and many other substances do not. And this allowed them to identify that the dioxins that led to the slaughter of the sheep and the ban on grazing within twenty kilometers of the industrial zone belong not so much to the emissions of furnace number 312 (being a high emitter, the dioxins that come from that furnace fall beyond twenty-kilometer distance) but the low-level emissions, those mostly contained in the electro-filters.

I'm convinced that the environmental question must be approached not only through the scientific aspect but also as a cultural question. We must also determine how aware the people are, because when you're inside of it you think everyone knows what you know, but that's not the case.

You're happy when you see that, little by little, you're able to convince the people who follow you. I see it as a service that I undertake for the community. In the beginning a lot of people said, "But what does she want, who is this Daniela Spera, does she want to run for office?" And so part of the work I've had to do over the years has also been to gain trust and credibility. And that's not a small thing, you know?

July 2, 2016

CHAPTER 4

Toxic Tales
Mapping Plurivocality, Mostly on Page

The stories that have been crafted about Taranto are so very many—too numerous to be adequately addressed in this book, and surely much greater in number when you read these lines than when I wrote them. In the past dozen years, as the debates about Ilva's future have raged on and new illnesses have been diagnosed, residents have also borne witness to a thriving narrative renaissance.[1] In the spirit of Laura Conti, contemporary authors, filmmakers, dramaturges, and visual artists from within the city and beyond seek to tell Taranto's many tales—largely of illness and ecological devastation but also of heroic histories and new beginnings. They do so in realist novels such as Cristina Zagaria's *Veleno* (2013, Poison), film projects like *Non perdono*, the hybrid documentary by Grace Zanotto and Roberto Marsella (2016, I do not forgive / Nonforgiveness), physically dynamic theater pieces such as Anna Dora Dorno and Nicola Pianzola's *Made in ILVA*, and more.

Through a vast range of storytelling practices and methods of circulation—including both mainstream publishers and free video-streaming websites—these contemporary descendants of Taras give voice to the living subjects of today's Taranto. Telling stories of singular protagonists and expansive communities, they explore what it means to have inherited stark legacies of risk and toxic embodiment, as well as rich cultural traditions and unscripted futures. In this they breathe new creative energy and a renewed sense of agency into a city that desperately needs it. What's more, they usher in a potentially vast audience to affective response and fuller cognition of an urgent ecological reality. Inviting readers and viewers into rich storyworlds, they alert us to the eco-catastrophe unfolding in Taranto, an occurrence that shares qualities with so many other potential and real catastrophes throughout the world, just as it is shaped by the particular coordinates of Taras's one-time home.

The wide array of creative narratives depicting the ongoing crisis in Taranto certainly outnumbers the scant but significant texts that recount the Seveso disaster. As a clearly marked moment in time, one with steps to be retraced, Seveso's ICMESA explosion inspired a small handful of linear texts dedicated to chronicling the events leading up to and then following it. Conti's *Una lepre con la faccia di bambina* stands alone as a fully developed creative narrative, at least one produced contemporaneously to and focused explicitly on the event's unfolding. In its near-solitary existence, and especially in the tepid response it received, Conti's work suggests that the community of Seveso was not particularly interested in having its story shared with a broader audience that may have already feared the contaminated area and its residents. And in its use of indirect micronarratives, an example of a more imaginative and nonrealist mode, it speaks to the fact that, as Conti underlines, expanded practices of communication and comprehension in and about Seveso were in fact gravely needed.

The ongoing and quotidian nature of Taranto's crisis makes for different kinds of stories and a much more robust output. While the residents of Seveso were loath for outsiders to know of the ICMESA explosion and its possible effects on their land and bodies (to say nothing of the material goods, such as furniture, that they produced and hoped to still sell), so very many in Taranto want for those beyond their community to understand the effects of Ilva's emissions on their land and bodies, to care, and to take action. There is a shared sentiment in Taranto, one of the southernmost cities in the southern region of Puglia, that their health and well-being matters less than the health and well-being of Italians farther north, a sentiment with deeply rooted cultural precedents dating back to Italy's Unification, if not before.

And yet, there is nothing mysterious about Taranto's eco-corporeal crisis. In our current age of advanced industry and environmental science, the facts are quite clear: the process of manufacturing steel releases a slew of quantifiable toxins into the air every single day, and these toxins cause a variety of illnesses. The result of all this for Taranto-based narrative texts is twofold. On one hand, a new normal has emerged: a daily and thus deflated awareness of everyday crisis to be incorporated as silent backdrop into texts focused on heists to pull off, romances to explore, family dynamics, and more. Examples include the drily comic *L'umano sistema fognario*

(Cosimo Argentina, 2014, The human sewage system); the *noir*-tinged *Nella perfida terra di Dio* (Omar Di Monopoli, 2017, In the nefarious land of God); and the thriller *Ombre sulla città perduta* (Silvano Trevisani, 2017, Shadows over the lost city). Notably, although these works avoid overt discussion of Ilva and its effects, they are all three quite dark in tone, as suggested by their titles, and recount crises of both spirit and employment (body, too: the first opens with the protagonist's mother dying from a chronic illness, surely a reference to the high levels of disease in the area).

On the other hand, many Taranto-based texts foreground the complicated effects of Ilva and its emissions on the greater community, inviting receptive audiences into a deeper place of knowledge and care. Some are overtly documentary in approach and offer chronological records of a given time period, while others weave references to nonhuman environment, human health, employment anxieties, and the polluting steelworks throughout fictional plotlines. While these recent Taranto texts are diverse, two common traits strongly emerge. The most intriguing for this study of narrative immersion and toxic transmutability is an attention to spatial mapping, to charting out the geographic coordinates of real-world Taranto such that far-flung readers and viewers are able to share a nearly corporeal sense of place and the matter inhabiting it. The second is a notable plurivocality, as many texts feature multiple narrating subjects, thus suggesting a horizontally diffuse expressive agency that is as inclusive as it is anti-authoritative. Both traits, mapping and plurivocality, underscore a desire for receptive audiences to understand Taranto's contemporary reality, and as such do tend to be presented in a realist mode. Where Laura Conti's adolescent protagonists step into the realms of imagination and role-play, sharing moments of make-believe in order to parse out the hazy details of a nationally recognized disaster, here, authors, artists, and directors tend toward realism in their urgency for audiences to know the concrete realities of an understudied case and place.

As scholars such as Raffaele Palumbo Mosca, Stefania Ricciardi, Luca Somigli, and others have explored, a so-called New Realism has emerged in Italian literature and film since the early 2000s. Maurizio Ferraris first introduced the term "New Realism" as a response to *pensiero debole,* or "weak thought," a philosophical approach advanced by Gianni Vattimo and others in the 1980s. *Pensiero debole* is a postmodern ethic emphasizing the process of interpretation, and thus the role of subjective experience, that is always at play in making sense of the world and conveying information through language. As Vattimo writes, "truth is the product of interpretation not because through its process one attains a direct grasp of truth (for example, where interpretation is taken as deciphering, unmasking, and so on), but because it is only in the process of interpretation, in the Aristotelian sense of *hermeneia,* expression, formulation, that truth is constituted."[2] *Pensiero debole* thus opens intentional space for the positive potential of subjectivity, for a truth that ultimately remains labile.

In a 2011 article in Italy's national *la Repubblica,* Ferraris heralds New Realism as the return of a philosophical and political approach to the world that is less interested in the openness of interpretation and, as he suggests, grounded in a firm ontology. Writing of the need to apply "greater attention to the external world," he articulates a notion that had already begun to be discussed by Italian scholars regarding literature and cinema in the mid-2000s.[3] Palumbo Mosca cites a 2006 issue of the journal *Allegoria,* entitled "Return to Reality? Narrative and Cinema at the End of Postmodernism," as the starting point for critically engaged discussion of a new or resurfaced realism in Italian literature and film. He also proposes, however, that Italian authors "began to show a renewed interest in history, as well as in social and political issues neglected by many of their immediate predecessors . . . as early as the late 1990s."[4]

While debates about the label New Realism continue to unfold, it is clear that Italian representational practices, especially but not only in literature and cinema, have been marked by close attention to the social and political precarity, economic instability, and environmental distress of recent decades. Reflecting the ideological motivations of their neorealist predecessors, as well as the naturalist formal tendencies of the literary *veristi* before them, many of today's Italian authors and filmmakers elect to portray postmillennial Italian life with a heavily realist gaze. The most well-known example of this is undoubtedly Roberto Saviano's 2006 novel *Gomorra,* followed by Matteo Garrone's 2008 film adaptation of the same title, which depict the brutal realities of life in the Vele housing project outside of Naples, so dominated by the Camorra's regime of organized crime.

Reflecting one strain of debates around the label New Realism, Pierpaolo Antonello reminds us in his work on the entwined *Gomorra* texts that, "as in any art form, there are formal and representational elements that cannot be neglected, and which are clearly present."[5] The same can and must be said for the Tarantine texts addressed here. While they are deeply engaged in conveying the difficult realities of life in the coastal city, they are not at all lacking in attention to form and artistry. Instead, they make ample use of the potential for representational media to communicate storyworld experience through various expressive and at times experimental means, even while maintaining a realist approach. I argue, in fact, that creative expressivity and attention to formal structure can be particularly useful in depicting a reality such as Taranto's, which is complicated by the imperceptibility of the toxic substances in question, as well as by the slowness of the economically and politically motivated social processes at play. As Raffaele Donnarumma notes, the latest return to realist portrayals of Italian society often reflects "the need to share a truth that surpasses the limits of what has empirically occurred."[6] Indeed, the many entwined truths of the Tarantine experience push storytellers and their audiences beyond traditional confines of witnessing and recounting into multidimensional and nearly multisensorial narratives.

To begin working on this project, I spent a year dedicated simply to reading and viewing what Taranto-based texts I could find. Through that immersive experience, I came to feel firsthand the power of entry into environmentally situated storyworlds. I began to think through Taranto—to traverse its streets, mourn for its sick children, relish its salty sea air, and wonder about its future with or without Ilva. In the summer of 2016 I visited Taranto for the first time, confronting the real city with its image in my head. What I found through long walks and longer conversations was in many ways the same Taranto I had already come to know, only spilling over with more stories and narrators than I had even imagined. It was as overwhelming as it was inspiring.

I arrived in Taranto from Rome late at night, midweek. As my second train of the long journey down slowly rolled into the city, industrial lights visible through the dark, I realized that I carried an excitement mixed with palpable apprehension. I was physically entering into a space and community with whom I already felt somehow bonded, having lived that space and those relationships through so many texts, but the place and people did not yet actually know *me*. The next morning, I embarked on an exploratory walk through Taranto's Città Vecchia (Old Town), a slip of an island flanked on both ends by bridges that divide the city's small bay into two halves. It was quieter than I had anticipated from my narrative wanderings, hardly another person in sight, and yet Taranto's physical shape, the curve of its coastline, was just as I knew it to be. As I looked out at the section of bay leading to the Ionian Sea, I almost felt Flavia Piccinni's teenage narrator pass by on her scooter, as she so often does in the novel *Adesso tienimi* (2009, Now hold me). Later, I couldn't stop myself from gasping with odd glee when I came across a simple drinker's tableau: a dilapidated office chair and spindly wooden side table set against an alley wall, an empty Raffo beer bottle perfectly perched. This was, I knew well, the locals' beer of choice.

What I found in the coming days continued to confirm my storyworld knowledge in ways both eerie and exciting while also, of course, surpassing it. I was profoundly struck by the oppressive weight of the air—a combination of midsummer heat, the city's slow pace, and the raw impact of seeing firsthand Taranto's environmental and health crises. During our shared wanderings, the folklorist and cultural guide Angelo Cannata repeatedly made offhand jokes about dioxin "getting to the heads" of local acquaintances. I caught myself wondering if it had gotten to me as well: by the end of my second day of walking and talking with the cultural advocates and environmental activists I had come to meet, I felt a guilty desire to board the next train out. The situation seemed so desperate, the contrast so acute between the reality in Taranto—where talk of Ilva's then still potential sale, gross environmental violations, and children's tumors passes daily across the lips of neighbors—and the rest of the country, taken up with other concerns.

I stayed on, of course, and as I continued to meet willing interlocutors and to let Taranto seep into my real-world consciousness, I discovered a still-expanding world of creative communication even more dynamic than I had first understood. I also heard heartening reports of active and affirmative response. I had gone to Taranto in particular to meet the aforementioned Daniela Spera, whom I felt I already knew through Cristina Zagaria's *Veleno*. Published in 2013, *Veleno* is a *romanzo-reportage*, a novel based on real-life characters and events (the book's complete title is *Veleno: La battaglia di una giovane donna nella città ostaggio dell'Ilva*, or Poison: A young woman's journey in the city held hostage by Ilva). A journalist by training who still writes for the national daily *la Repubblica*, Zagaria has written many books in this style, all focused on female protagonists fighting against challenging circumstances. *Veleno* follows Daniela Spera in the third person during the years 2010–12, as she gathers crucial evidence to build a case against Ilva and eventually speak before European Union courts. In the novel, as in life, she works from the ground up with a passion and dedication Zagaria and others have likened to that of the American grassroots activist Erin Brockovich, about whom Steven Soderbergh made a popular film in 2000.

As Zagaria pointed out to me during an afternoon in Naples just prior to my Taranto visit, Daniela's last name, Spera, means "she hopes" (from the Italian verb *sperare*). It is this hope, along with her training in empirical research as a chemist, that nurtures her persistent dedication to speaking up for the well-being of her community. It is also her belief in the importance of individual stories. Zagaria explained: "Over the years Daniela had met various people who told her their stories, and she was compiling a file. Her plan was to go to the European courts, so she was collecting all their stories—and then I started to go with her to meet all these people and they told *us* their stories."[7] Those stories found their way into Zagaria's novel, and into Daniela's testimony when she and colleagues from the local environmentalist group Legamjonici spoke before EU magistrates in 2013. The stories underline that the crisis in Taranto is about more than just data; rather, it involves thousands of individual lives and family struggles that share a common thread. When grouped together, such stories—of longtime factory workers and young children alike—trace out a "common path" rather than standing alone as "those pitiful symbolic cases."[8]

The fact that residents were willing to share their stories, first with Spera and then with Zagaria too, speaks to something that I have encountered in my Taranto wanderings as well, and touch upon above: as opposed to the residents of Seveso, many Tarantini are eager to be heard, so ready to narrate their tales of polluted environment and ill health in the hope that others might know and take action. Over the course of a late-morning coffee that somehow turned into dinner, Spera shared with me her own story. She described how her work as a pharmacist, in which she hears daily of local illness, led her to become one of Taranto's most outspoken environmental health advocates. She told me about the chemical makeup of dioxin and why long-term studies are more revealing than those daily or even monthly peaks in reported emissions, about her mother's recent sudden death from an aggressive cancer, and about her attempts to communicate Taranto's realities on levels both local and international through public lectures, government hearings, and, increasingly, writing. When I asked her what it was like to have circulated the world in written story form thanks to Zagaria's novel, she was less forthcoming, almost embarrassed. But when I asked her about reaction to the work, she beamed, telling me: "As a novel . . . it created an awareness for many young people. In fact, there were a few girls who wrote to me, saying: I study at the university in Pisa now—I chose environmental science. I'll tell you something more: I'm going to graduate from Pisa and then I'll go back to Taranto, because I can go back to Taranto possessing *knowledge*." A knowledge, one hopes, that might lead to change.

MAPPING

On the act of walking in the city, Michel de Certeau writes: "In modern Athens, the vehicles of mass transportation are called *metaphorai*. To go to work or come home, one takes a 'metaphor'—a bus or a train. Stories could also take this noble name: every day, they traverse and organize places; they select and link them together; they make sentences and itineraries out of them. They are spatial trajectories."[9] This reading of *metaphorai* is particularly applicable to three aforementioned Taranto novels: *Adesso tienimi* by Flavia Piccinni, *Veleno* by Cristina Zagaria, and *Vicolo dell'acciaio* (2010, Steel way) by Cosimo Argentina. All three are set in contemporary Taranto and marked by a heavy use of spatial mapping. Repeatedly listing street names and town squares, describing the distances between locations and the feeling of traveling between them, they trace out spatial trajectories for primary human protagonists within the texts, as well as for readers. At the same time, they chart the course of the toxic matter released in Ilva's emissions that does those humans harm. Certeau explains that "every story is a travel story . . . a spatial practice," in that it takes protagonists and thus readers from one location to the next, charting a path between them. In fact, he writes, stories "organize walks. They make the journey, before or during the time the feet perform it."[10] In the case of present-day Taranto, the stories of Piccinni, Zagaria, and Argentina describe and define the spatial paths for the feet (or wheels) of human subjects but also for the other forms of matter that inevitably accompany them.

Flavia Piccinni's debut novel, *Adesso tienimi*, was published by Fazi editore in 2007, the same year in which ARPA Puglia reported Ilva's dioxin emissions to be twenty-seven times the European limit.[11] The novel is narrated by seventeen-year-old Martina as she stumbles through her final year of high school after the suicide of her abusive lover, who was also her teacher. Martina's is a bleak tale that grows steadily darker. As the story unfolds, she delves further into a state of despondence, mourning in secret, and the usual uncertainties that come with the end of high school and adolescence bear down with particularly pronounced weight in a Taranto marked by workers' strikes, religious processionals, "scarcity" and "shortage," and a chemically pink sky.[12] What's more, her love affair eventually manifests in memory as another form of Taranto's toxicity, to employ a common and distinctly nonmaterial discursive use of the term. Like local dioxin, the relationship is a harmful something imposed on her (the deceased lover was its aggressive initiator), but which she can no longer imagine living without. In its focus on female adolescence in an Italian steel town, and the connection it draws between polluted environments and troubled human behaviors, *Adesso tienimi* holds clear parallels with Silvia Avallone's *Acciaio* (2010, published in translation as *Swimming to Elba*, 2012). Enrico Cesaretti writes that Avallone's text, the title of which literally means "steel," "illustrates the difficulty (if not the impossibility) of isolating the health or illness of geographic spaces from the health or illness of existential spaces."[13] The same can certainly be said of *Adesso tienimi*.

The linear story takes place over the course of a few short months and is full of spatial movement as Martina brings readers along from one location to the next. Every few paragraphs, it seems, we are somewhere else: her bedroom, a street corner, a restaurant, a friend's home, in rare flashback at her lover's beachfront getaway, or, more often than not, speeding along roads on the back of someone's scooter. As she travels, she maps out the city's streets and neighborhoods, references by brand name the favored beer and cigarettes of its teens and workers, and traces the wide spread of its industrial emissions, from factory to sea to skin. In this, as well as its insistence on present moment and forward movement through both verbal tense and narrative action, *Adesso tienimi* embodies the speed that Gianluigi Simonetti and others call so emblematic of contemporary realist Italian narrative.[14] Martina, and by extension we, are always wandering in this novel. The spatial trajectory remains strictly within the confines of a well-mapped Taranto, however, as reinforced by a steady practice of reciting place-names: via Crispi; the Salinella neighborhood; viale Magna Grecia; the Ponte Girevole rotating bridge that crosses the Mar Piccolo into the Città Vecchia; Ugo Bari (a popular trattoria in the historic city center that has since closed); the Piazza della Vittoria; Ilva.

Martina holds her own emotional state as an explicit point of narrative focus. The city itself is implicit as her primary nonverbal interlocutor and affective influence, however, as established by *Adesso tienimi*'s opening lines: "I was born in Taranto. A debt of 500 million and 90.3% of the dioxin killing Italy. I live at via Cagliari 32/A, in a nice little white building with a doorknob made of rusty hammered steel. I smoke two packs of blue Chesterfields a day, eat only sugar-free gummy candies and cheese popcorn. In my free time I watch television or cry."[15] Thus introducing herself to readers, Martina self-identifies first and foremost as being *of* Taranto, which she defines here exclusively in terms of economic and eco-corporeal ruin. She then draws a direct correlation between those negative states and the trappings of her life. The steel doorknob to her home, presumably made from Ilva's products, is rusted; her time is spent either in physical manifestation of emotional pain or in a state of distraction from real life (she prefers animated children's programing above all); and her own body is sustained by a diet largely devoid of nutrients. Living on artificially sweetened gummy candies, Diet Coke, popcorn, and dozens of cigarettes a day, Martina becomes the human embodiment of Taranto itself, surviving almost entirely on fumes.

She is calmly honest about the city's pollution throughout the course of the novel, interweaving her study of place and self with references to Ilva's toxic emissions as they pervade both natural environment and human body. She frequently notes the color of the sky when describing her wanderings, stopping midnarrative to explain: "The sky in Taranto is never blue, even when it seems like it is. It doesn't depend so much on the geographic position. But on the intensity of the fumes. Streaked with red during the day, gold at night."[16] Martina is clear-eyed and unremorseful in her assessment of Taranto's supranatural sky, which has presumably been polluted her whole life. In fact, she finds the chemically modified view beautiful, and again closely tied to her sense of self. In her deepest state of despair at novel's end, Martina pauses to list the most pertinent things in her seventeen-year-old life. She thinks of much-desired jewelry ordered on the website eBay, of her upcoming high school graduation, of friends soon to move away, of favorite foods, and "of Ilva, which colors the sky and makes it seem more beautiful to me."[17] Ilva is part of her city, and so it is part of Martina. While this sentiment can be interpreted figuratively, textual passages in the book suggest that it is also to be taken literally, in line with what Past might call an "epidermic" understanding.[18]

Appreciative of the aesthetic effects of Ilva's emissions, Martina is also deeply aware of their power to enter and alter human bodies. Riding past the Tamburi neighborhood one afternoon, she reflects on "the highest percentage of lung cancer deaths in the peninsula . . . the sea swallowed up by mercury, on which fish are drugging themselves[,] . . . tomatoes on the vine, and sheets, which are already colored with red dust. Ilva red."[19] It is in Martina's easy cognitive transition from toxins in the air to toxins in the lungs that Piccinni's text slides most fully into a discourse of embodiment. The red steel dust, on which dioxin and other compounds are carried throughout the city, merges not only with air but also with plants on the vine, water and fish in the sea, and residents' own bodies. The vast web of material connectedness Martina describes confirms that, as Nancy Tuana writes, "the boundaries between our flesh and the flesh of the world we are of and in are porous. While that porosity is what allows us to flourish—as we breathe in the oxygen we need to survive and metabolize the nutrients out of which our flesh emerges—[it] often does not discriminate against that which can kill us."[20] What's more, that web underlines that we cannot address issues of human health (like lung cancer deaths) without considering the health of the nonhuman, and that we cannot attempt to remediate the natural environment without considering the human sources (like Ilva) that do it harm. This attention to the interconnectedness of environment, industry, and health again speaks to the depth of Piccinni's narrative, an intimate coming-of-age tale that doubles as rallying cry for change.

Martina's message is loudly echoed in the aforementioned *Veleno*, another text that sees its protagonist repeatedly traverse the city, as she goes from home to pharmacy to the heavily afflicted Tamburi neighborhood, walking the streets to speak with residents. In this novel, published three years after *Adesso tienimi*, we read: "Every dioxin leaves an imprint that links back to whoever produced it. . . . It's as though little murderous hands were placing themselves on us, touching us, each choosing a part of our bodies . . . lungs, head, ovaries, skin, throat . . . in order to then devour it, leaving only their little imprints."[21] Writing in the fictionalized voice of the real-life Daniela Spera, the author Cristina Zagaria traces Spera's daily movements, just as she traces dioxin's path from industrial producer to human body. She underlines the toxin's dependent origin (it was produced by some other agent or process) just as she confirms both its destructive agency and our own human mutability.

Midway through the text, Daniela and friends take a much-needed break from daily routines to go to the beach. Of their drive away from the city, we read: "They brush past the cemetery, go around the traffic circle, and there they are on Route 106: Ilva to the right, Eni to the left[,] . . . quickly leaving behind the Church of the Cross, Punta Rondinella, the industrial docks, the shipping containers."[22] As the group exits Taranto, the city's threshold is marked by the cemetery on one side of the road, Ilva and the Eni oil refinery on the other, church and symbols of industrial trade in between. It's all right there: production and distribution, life and death. Through this and similar descriptions of human movement through more-than-human geography, *Veleno,* like *Adesso tienimi,* invites readers to create a mental map of Taranto, just as it brings us into deeper awareness of an unsettling relational ontology in which both human bodies and dioxin serve as the link between burial ground and steelworks.

In a final example from *Adesso tienimi,* protagonist Martina further confirms her understanding of toxic embodiment via spatial movement and porosity, once again with unnerving calm, in a rare moment of connection with her mother, Adriana. In this scene the two women convene in Martina's bedroom on Palm Sunday, the last Sunday before Easter in Christian tradition, after Adriana has returned from church services with a palm frond for her daughter. Describing the exchange, Martina notes: "Adriana puts the palm leaf on my desk. She sits on the edge of the bed and says that this morning, when she went out, she couldn't breathe. She explains that there was this sort of violet cloud that, they told her, came from Ilva. When she says that the poisons are coming into our houses I nod. I would like to tell her that it's so true, that the poisons are near us and inside of us, but then she smiles at me and points to the palm."[23] What is for Adriana a rumored exterior presence slowly encroaching on the domestic sphere, is for Martina already an interior reality, moving through her body like—and indeed along with—the cigarette smoke that she continuously inhales.

The difference between mother and daughter here suggests a generational divide regarding the primacy of toxic experience and awareness, a divide that Martina silently maintains by not giving voice to the poisons already inside her, inside them. Readers note that she is silenced by her mother's smile and indication of the palm leaf, a religious symbol of the spirit's victory over the flesh. If her reflective narrative tells us anything, it is that for Martina spirit and flesh cannot be so easily separated, either from each other or from the world that surrounds them. Through Martina's story, Piccinni does not propose a clear path of repair for spirit, flesh, or environment in contemporary Taranto. What she does offer, though, by way of careful attention to a very real contemporary crisis, is an understanding that industry, environment, and health, both physical and beyond, are deeply intermeshed. By grounding her text in spatial mapping, she invites readers to imagine what it might be to perform the physical journey that is life in Taranto and, in a sense, to experience that mesh firsthand.

I highlight *Adesso tienimi* for the ways in which it seamlessly incorporates attention to toxic embodiment, transcorporeality, and place into a mainstream coming-of-age novel, thus underlining both the vast reach of Taranto's toxicity and the potential for storyworld experience to increase eco-corporeal awareness far beyond the bounds of traditional nature writing or declaredly environmentalist texts. I also highlight *Adesso tienimi* as one of the titles to have achieved greatest commercial success from among the many Taranto-based narratives published in recent years. Recipient of the Readers' Prize from the Società Lucchese dei lettori in 2008, the novel's first edition has sold well over five thousand copies, and in 2019 it was reissued by the Puglia-based publisher TerraRossa (Piccinni has also been awarded the Campiello Giovani and Essere Donna Oggi prizes). A 2018 review of the novel by Carlo Caprino notes that its depiction of "a particular and universal Taranto" still rings true ten years after initial publication.[24] The arts and culture blogger, based in the Taranto-proximate community of Grottaglie, goes on to write, "More and better than so many sociological analyses and journalistic inquiries . . . Flavia Piccinni recounts Taranto and the *Tarantini*."[25]

Alongside Piccinni and Zagaria's novels, the work of Cosimo Argentina merits special attention when it comes to the practice of spatially mapping contemporary Taranto. As Mario Desiati suggests, Argentina might just be *the* Tarantine writer.[26] Like Piccinni, he was born in the coastal city, and he continues to return to it through his writing. Thus far he has penned the novels *Il cadetto* (1999, The cadet); *Cuore di Cuoio* (2004, Leather heart); *Maschio adulto solitario* (2008, Single adult male); *Vicolo dell'acciaio* (2010, Steel way); *Per sempre carnivori* (2013, Carnivores forever); and the aforementioned *L'umano sistema fognario* (2014, The human sewage system); along with the essayistic *Nud'e cruda: Taranto mon amour* (2006, Naked and raw: Taranto, *mon amour*). All contribute to what Angelo Prudenzano declared in 2009 to be an "authentic *nouvelle vague*" emerging in Taranto and the Greater Salento region, one dedicated to proudly dispelling the "image sold to tourists" of tranquil beaches, folklore, and home cooking.[27] The first title begins in Taranto but quickly sees its narrator move on to other locales, the second is set in Taranto of the 1970s, and the third takes place in surrounding provinces. It is with *Maschio adulto solitario* that Argentina begins to explore Taranto's more recent history, in a fictional narrative that spans from the 1970s to the 1990s. Here he introduces the themes of factory life and toxic emissions that are most fully elaborated upon two years later with the present-day *Vicolo dell'acciaio,* then alluded to with the terminal illness that introduces *L'umano sistema fognario* and sets its plot in motion.

Like Piccinni's novel, *Vicolo dell'acciaio* features an angst-ridden teenage protagonist, Mino, who is still sorting out his relationships to family, future, his girlfriend, and Taranto itself. It is more overtly focused on Ilva and its damaging effects than is *Adesso tienimi:* Mino's father works at the steel mill, as do most of the adult men in his apartment building, number 75, and along his whole street, via Calabria. It is from this concentration of steelworkers that the novel gets its name. Bearing the poetically blunt dedication *ai fottuti* (a softened translation would read "to the damned"), *Vicolo dell'acciaio* does not beat around the bush when it comes to pervasive illness in Taranto. After three succinct lines of dialect-tinged dialogue announcing the death of a neighbor ("Cit'ammuert!" / "Ce stè ddic?" "Ianegle Cite . . . du 'u quart' pian'?"), the novel's opening paragraph reads: "Via Calabria 75, an apartment building in which 90 percent of the heads of families get their kicks at the steelworks. From the first to the seventh floor we're all encrustations of the great maternal pipe and our skin smells like the steelworks' iron gold."[28] Three pages later: "Here in our building everyone dies from lung cancer, the winning track record is ours. Per family, we have more benzene, carcinogenic dust, dioxin, aromatic hydrocarbons, and saturated gasses in our bodies than I don't even know what."[29] The prevalence of terminal illness is staggering in Mino's Taranto, and yet, as he explains, his generation will continue working for Ilva (or Italsider, as it is often still called by locals both in the book and in real life), unless they find a way to bypass their common destiny. When the novel begins, Mino is half-heartedly attempting to do just that, studying for exams with plans to transfer to the university in Bari. As the novel progresses, and Mino's friends and family continue to be struck by illness and death, its primary narrative conflict emerges: will he join his girlfriend and local environmentalists as they fight against Ilva, or will he abandon his studies and join the ranks of the "true 75s" himself?

Where *Adesso tienimi* and *Veleno* are both marked by an energetic mapping, one that follows the moving bodies of its protagonists to chart both the city and the dioxin-tinged toxicity moving through it, *Vicolo dell'acciaio* features a mapping of a different sort. In this work, Taranto and its residents are already so inundated with Ilva's toxic matter as to be one large animate mesh of steelworks, city, and people, simultaneously vibrant and necrotic. As such, Argentina's narrative is less motivated than the other two texts to guide readers dynamically through the city in explication of transcorporeal flows; every *thing* is already so clearly every *where*. Although his protagonist does still travel through the city at times, he is more stationary than the two protagonists discussed above, and deeply tied to the very local community of his apartment building and immediate neighborhood.

Mino still maps the city verbally, however, obsessively listing street names, building addresses, and neighborhoods any time he introduces a new character or contemplates his evening plans. He does this so frequently, in fact, as to allow readers to create our own free-form mental maps of the city, with via Calabria intersecting via Polibio in central Taranto, neither as close to Ilva as the Tamburi neighborhood nor as close to the city's stately central piazza as via Regina Elena, where the aforementioned environmentalist group gathers. Mino's impulse toward place-based identification in *Vicolo dell'acciaio* lies in establishing affiliative networks between people in light of the places to which they "belong." Physical proximity creates both kinship and knowledge structures in the novel: as Mino explains mid-text, "every time one of us leaves the neighborhood, he's disoriented."[30] For this reason, "true via Calabrias at the most will go listen to some music at the Ramblas," which, he lets readers know, "is a club in the Canale zone."[31]

This frequent naming serves as a sorting mechanism to establish ties but also to identify the concentration of steelworkers in any given block, as well as proximity to Ilva itself. A friend's uncle with a year to live is referred to as "a via Salinella," another acquaintance is identified simply as "a via Dante," the festively named Johnny Lupino is, more precisely, "a via Polibio, number 70," and so on.[32] Proximity to Ilva or to a great concentration of its intoxicated workers legitimates certain types of speech about the steelworks and its possible futures, with Mino ascribing greatest authority to those who live "on the front lines."[33] It also lends itself to a certain mythologization, made evident through a lexicon of battle, honor, and glory used to describe the men who work at Ilva, even when it comes to their illnesses (Mino notes reverentially of one longtime steelworker, "he's toxic even from a distance").[34] Mino speaks frequently of the workers as being in the "trenches," and he calls his father "the General" with great admiration. While this nickname originally derives from a humorous incident during Mino's infancy, he regularly describes his father's time on the job as akin to being on a battlefield. In one instance he imagines the General working the steel "almost like an ancient noble warrior forging his own weapons."[35]

This description of worker as warrior, molding steel that will protect him in battle, offers up a bitter irony, as the production of that steel will, by novel's end, have killed Mino's father. Mino knows this, of course, and in a less poetic moment remarks: "At this hour my father, the General, is there beyond the two bridges. From where I stand I can see the power of Italsider's smoke. I can see how the monster flexes its muscles and ridicules us."[36] With this reference to Italsider / Ilva's power, and with its figuration as monster (a figuration not uncommon in local speech, I have found), we enter most fully into the discourse of animacy that permeates *Vicolo dell'acciaio*. As Mel Chen writes, in linguistic terms animacy refers to "the quality of liveliness, sentience or humanness of a noun or noun phrase," but it is also much more, slipping beyond "language's bounds" just as language slips "out of animacy's bounds."[37] Animacy is also, in Chen's formulation, a permeable notion by which scales of "aliveness" or "deadness" are measured through hierarchizations of matter that are inherently imbricated with questions of race, class, gender, and disability. This hierarchy is for Chen "naturally also an ontology of affect: for animacy hierarchies are precisely about things which can or cannot affect—or be affected by—which other things within a specific scheme of possible action."[38]

Vicolo dell'acciaio asks readers to consider the animacy of Ilva as a singular (and yet transmutable) entity against, alongside, and entwined with the animacy of human subjects, much like *Adesso tienimi* and *Veleno* ask us to do the same for dioxin. Following Chen's formulation, in *Vicolo dell'acciaio*'s animacy hierarchy Ilva tends to win, often exhibiting more liveliness than the creatures in its midst, and certainly wielding a greater affective power than they do. The novel explores Ilva's animacy by positing the steelworks as a monster, as referenced above, but also as an overbearing human subject. Mino muses: "Italsider occupies an apartment in this neighborhood. Unfortunately, the apartment's two times the size of Taranto. From its domicile, a '*centomila vani e accessori*,' the steelworks couldn't care less about the emissions ban, the ban on acts of emulation, and all the condo association rules you can think of."[39] Here in the guise of a nightmarish neighbor, Ilva is ascribed not just a private domicile but feelings: it couldn't *care* less about the rules of community conduct.[40]

It is not just the steelworks that possesses a creaturely liveliness and affect in this text but also the city itself. In a rare instance in which Mino does walk through Taranto, he likens it to a human or animal, just as he once more describes the steelworks as a powerful monster. Charting a course similar to one that *Adesso tienimi*'s Martina might take (if you look at an actual map of Taranto, you will find that the fictional characters' stated addresses are separated by just a few blocks), he explains: "I take via Giovan Giovine and I come out onto via Dante . . . from Corso Umberto, I end up along the Canal. . . . I cross over the bridge and I'm in the Città Vecchia. I lick the 'Arabian city' along the inside of its flank and then I'm in the Tamburi neighborhood. From there I observe the monster, the steel hell."[41] Like an animal, Taranto here possesses a flank (the original *fianco* might also be translated as "hip"), one that an ambulating Mino, in a surprisingly sensual linguistic turn, "licks" or perhaps "kisses." Taranto's animal liveliness and sensuality are later presented as sadly diminished, however, when Mino describes the city as a used-up prostitute: "Via D'Aquino gives a sense of the state in which the city finds itself. A whore at the end of the line, this is what Taranto has become. A whore through and through, but with aching feet and her calves a mess."[42] Here Taranto is simultaneously anthropomorphized and objectified, lowered in the novel's animacy hierarchy (notably, through attributions of gender, class, and sexuality) and both used and acted upon by the steel-producing "monster."

As Iovino reminds readers in her work on the "porosity" of Naples, another picturesque southern Italian city marked by toxic tales, cities are indeed animate, and as permeable as human bodies: "As bodies are what they are via their permeable boundaries (membranes that cause the flows of energy and matter), so, too, bigger entities and formations follow the same dynamics. A city, for example, is a porous body inhabited by other porous bodies, a mineral-vegetable-animal aggregate of porous bodies."[43] Described as a living creature unto itself in Argentina's text, Taranto is in fact doubly animate, permeated by the porous bodies of its struggling residents.

In an additional expression of such permeability, as well as a further embrace of the aforementioned (zoo-)anthropomorphism, Taranto also *breathes* in the text, an action that seems to both sustain and diminish life, and that again recalls the dioxin-tinged transcorporeality of *Adesso tienimi* and *Veleno*. When Mino first speaks of his mother in the novel, he explains that although she was born in northern Italy, she is now fully of and in Taranto, just as Taranto is now fully of and in her. "But now she is in Taranto's breath," he explains. "She is in the pre-equatorial breath; her lungs travel along the Lecce to Battipaglia tracks . . . at this point she's one of us."[44] Not only does Mino once again ascribe creaturely animacy to the city, he also defies reader expectation by first establishing that *she* is in *it* (we are used to expressions of a place becoming "part" of a person, but not the other way around). He then complicates things, however, by transitioning from Taranto's act of breath, which presumably necessitates lungs, to an image of his mother's lungs moving along (and thus breathing) the local train route. In this quick slide from place breathing person, to person breathing place, Mino offers a compelling depiction of enmeshment between the two that speaks as much to affect as it does to transcorporeal flows. His mother feels an emotional connection to Taranto, but she is physically connected to it as well, through all the airborne matter that flows into her body and all the air-borne matter that flows out of it.

Later, describing their shared family home, Mino notes: "The fumes from Ilva enter into the kitchen, the living room, the bathroom. We inhale dioxin in the form of armed silences."[45] Much like the shared breath of city and its resident in the passage cited just above, the conduit for passage and transference in this description is also air. Fumes (gasses, smoke, vapors) are of course not only air but instead a combination of matter flowing freely through the environment by way of that primary, invisible, and universalizing element that knows no bounds. By reminding readers that human bodies contribute to the passage of matter, *Vicolo dell'acciaio* underlines the ways in which we are active, if not willing, participants in dioxin's flows.[46]

I cite a final example, when Mino visits his girlfriend's sick father, called Trottola (whirligig): "I go into the room of the damned and my nostrils fill up with the odor that Trottola is beginning to emanate. *Fizzo* of infected ulcers . . . His body is just a sack of shit at this point; it has been for at least six months. There was just this breath in there, like a tiny drool of life."[47] A lingering sign of his animacy, the air that comes from Trottola's body in the form of breath and other gasses is also the strongest sign of his illness. This, more than anything, might serve as an apt symbol for the complicated state of being in present-day Taranto: the very breath that circulates within and between creaturely bodies and the city itself—carrying dioxin in its mineral particles, along with the lingering spirit of Taras, Falanto, and their dolphin saviors, uniquely able to navigate all sorts of liveliness.

In all of the texts I have outlined above, a protagonist addresses environmental toxicity, acknowledges the flow of toxic matter through nonhuman nature into creaturely bodies, and leads readers through the city and its outskirts, casually listing locations as though we, too, should be familiar with them. By reinforcing both our imagined intimacy with Taranto and our potential real-life disorientation as outsiders, this spatial mapping encourages readers to consider Taranto as actual physical place, just as it requires us to remain active interlocutors with the text. Not only are we following a story's plot progression and character development, but we are also working to piece together a geographical model and understanding of spatial relationships in a way that roots our knowledge of these place-based texts firmly in the physical. Through such a process, these diverse narratives encourage readers to cognitively acknowledge Taranto as a real place and thus, hopefully, to care about the actual struggles of its residents and more-than-human environment. At the same time, by mapping the flow of dioxin and other toxic matter along with the movements of their protagonists, these narratives remind us of our own mutable physical beings, and of the porosity of the world at large.

PLURIVOCALITY

As a coda of sorts to the somber texts and themes discussed up until now, this chapter's final pages are dedicated to another significant trend in the Taranto-based texts that have emerged in recent years: an emphasis on multiplicity of narrative voice. From documentary films like *Buongiorno Taranto* (Paolo Pisanelli, 2014) and the experimental *Non perdono* (Grace Zanotto and Roberto Marsella, 2016, I do not forgive / Nonforgiveness), to the story collections *Ilva Football Club* (Fulvio Colucci and Lorenzo D'Alò, 2016) and *Sognando nuvole bianche* (edited by Simone Daini, 2012, Dreaming of white clouds), to the community-minded projects of the arts collective Ammostro, many of Taranto's tales refuse a singular narrative voice. Instead, they give space for many narrators to share their stories, embracing a Bakhtinian sort of polyphony in which multiple voices emerge within a single text. In this they underline the fact that Taranto is a community made up of complex individuals who refuse to be condensed into a singular image, despite shared experiences of illness or risk. At the same time, they work to counter the legitimacy of a singular authoritative narrative from Ilva executives and government officials, who argue that Taranto is doing fine—but only if the steelworks continues to produce.

By offering nuanced stories of individual experience that together paint a portrait of both suffering and resistance, the multiple narrators of the aforementioned texts participate in a counterhegemonic storytelling. To cite again Marco Armiero's work on the "guerrilla narratives" of Italy's Campania region, long battling a waste crisis of epic proportions, "narrating means producing counter-narratives, because environmental injustice is not only imposed with tanks and truncheons but also with a narrative that eradicates any possible alternative, that requires an official truth."[48] By producing not only counternarratives but plurivocal counternarratives, authors, filmmakers, and artists add complexity, intimacy and a multiaxial richness to our understanding of Taranto's ongoing crisis, just as they work to actively counter dominant narratives created by state and industry.

Such action is often so necessary in cases like Taranto's, when national economic interest may dictate a mainstream portrayal of events. While the Italian government has recently begun to acknowledge the environmental injustice in Taranto, both the state and Ilva (whether under Riva family executives or ArcelorMittal) have long downplayed reports of toxicity and obfuscated data on health and mortality as they promote the steelworks' national and regional economic importance.[49] By privileging and personalizing experiences of ongoing environmental illness, many of the authors and artists crafting contemporary Tarantine storyworlds instead propose that employment at Ilva is no longer worth the grave damage the steelworks has caused to environmental and human health, thus upsetting the long-upheld official narratives that Taranto needs Ilva to survive and that employment alone is enough to equal well-being. Gaetano Colella's play *Capatosta* (2014), for example, addresses these contrasting narratives through lengthy dialogue between a veteran Ilva worker and his new young colleague as they go about their workday. Ruefully echoing the message of their employer, the older man explains early on: "The steel that we make is much more important than all of our lives put together: mine, yours, your father's. Everyone's."[50] Yet in sharing the lives of these two Tarantine characters through theatrical representation, Colella disproves that message, privileging the telling of stories over their steelworks setting, which serves as mere backdrop for intimate exchange.

I will address the aforementioned films and arts collective in chapters 5 and 6, respectively, and focus here on Giuliano Foschini's *Quindici passi* (2009, Fifteen steps). Based in Bari, Puglia's capital city, Foschini has long served as a regional correspondent for the national daily *la Repubblica*, much like *Veleno*'s author, Cristina Zagaria. His reporting tends toward environmental and social justice causes, both heavily present in *Quindici passi*. With this first book (he has since coauthored two more) Foschini offers a nonfiction documentary text so attuned to a variety of personal narratives as to be a short story anthology as much as a journalistic chronicle.[51]

An investigation of Ilva, its emissions and human health, *Quindici passi* is also an ode to the many voices, and their many stories, that together craft the reality beyond Ilva's factory walls. In its rhythmic continual rotation of individual story, Foschini's book serves as a form of testimonial realism, in which he seeks to "gather interviews, respecting the individuality and voice of his interlocutors" while still allowing space for his own reflections.[52] Uniting as many voices as possible, Foschini offers a nuanced portrait of Taranto less concerned with one authoritative version of events as with "the necessity to say a truth that goes beyond the limits of the empirically occurred."[53]

As a structured accumulation of actual first-person voices, *Quindici passi* engages in a reality grounded in both multiplicity and commonality of experience. While each speaker's story and voice are unique, they are united in relationship to local environmental toxicity, as well as by Foschini's authorial ordering principles. By evenly interspersing his own reflections with those of new interlocutors and allowing each person's story to fill the same approximate space on a page, Foschini makes clear just how many diverse members of Taranto's community are affected by exposure to toxic substances such as dioxin. Like the novels addressed earlier in this chapter, *Quindici passi* is deeply committed to ontological narrative, disseminating knowledge through a storytelling practice guided by clear-eyed assessment of Taranto's ongoing ecological and human health disaster.

Published two years after Piccinni's *Adesso tienimi*, Giuliano Foschini's *Quindici passi* offers a more direct examination of human and environmental health in Taranto and, subsequently, a less complete picture of the city whole. A reportage structured around a collection of individual stories, the book's primary narrative plane is recounted in Foschini's own casually journalistic voice and describes a conference from late 2008 at Taranto's Testa hospital. Hosted by the aforementioned environmental group ARPA Puglia, the two-day conference was dedicated to the interchange between industry, toxicity, and illness. As he makes his way between presentations, Foschini reflects on statistical information shared by ARPA's officers and records testimony from medical investigators, as well as from community residents who have lost family members to various illnesses. He also thinks back on previous encounters with others attending the conference, like the aforementioned Angelo Fornaro, whose entire flock of sheep had to be killed after chemical analysis confirmed dangerous levels of dioxin in their bodies in 2008.

Throughout, the book is bracketed by citations from articles in newspapers such as *la Repubblica,* as well as from pediatric oncologists, factory workers, and politicians both local and national. It is further interspersed with text from letters written by concerned area schoolchildren, which can also be found in the aforementioned *Sognando nuvole bianche.* Finally, each movement of the text, from one part of the conference, its themes and stories, to the next, is linked by a common metaphor: Taranto as volcano. Foschini writes in the book's opening chapter: "Craters of cement, magma, and unemployment insurance, Taranto is a volcano. Active."[54] Like a volcano, he explains, the city is bubbling over with a devastating substance, ready to blow at any time. And as in the geographic area surrounding a volcano, Taranto's sky is full of fine particulate matter that makes for particularly striking sunsets, of the kind described in *Adesso tienimi.*

The title of Foschini's book refers to the distance between the Ilva steelworks and the nearest homes in the Tamburi neighborhood, the same distance between Ilva and the San Brunone cemetery. Like everything reported in this nonfiction text, that distance, while approximate, is true: both homes and burial ground rest just beyond the steelworks' border. Linking two very different types of "resting place" through shared proximity to Ilva, Foschini thus acknowledges the plant as the primary source of livelihood for so many Tamburi residents, at the same time that he implicates it in their disproportionately high deaths. Like the authors addressed above, Foschini is well aware of Taranto's economic dependence on Ilva and of the ways in which the plant's presence seeps into so many residents' sense of self and place. Again like those authors, he is similarly attentive to the ways in which its toxic compounds enter their bodies.

The first mini narrative portrait in the text features fourteen-year-old Luca, who suffers the sort of throat cancer that typically targets mature long-term smokers. By way of Luca's story Foschini also touches on that of Maria, who died from pancreatic cancer at forty-nine, while steadily weaving in statistical information shared by Alessandro, a fellow conference participant. This chapter, "Scusi mi fa accendere" (Excuse me, do you have a light), begins when Alessandro asks Foschini for a lighter, although neither of the two smokes. The writer understands that Alessandro's question "had to do with an allusion, that there was something behind it and that something was in some way connected to the city, to the color of the sky, to the port, the smokestacks. In essence to the volcano."[55]

Again as in Piccinni's text, cigarettes become a recurring motif in *Quindici passi,* since "in essence, there are cigarettes in Taranto's air."[56] For a period of time in 2004, Foschini writes, sixty-seven nanograms of dioxin could be found per cubic meter of air in Taranto, as though everyone in the region were in fact smoking 128 cigarettes a day.[57] This figure makes fictional Martina's few dozen daily cigarettes seem rather paltry while again underlining the connection between industry, environment, and bodily health. In its awareness of embodiment *Quindici passi* also engages questions of futurity, recognizing that current states of pollution and toxicity have long-term effects, in a way that *Adesso tienimi,* with its teenage immediacy and angst, does not. Midway through a chapter revolving around stories of childhood mortality, Foschini writes: "Dioxin wasn't just in the air. Or perhaps in sheep's meat or cheese. Dioxin had even made its way into the breasts of Tarantine women."[58] Then, in a later chapter focused on the necessity of work, he shares an interview with longtime Ilva employee Mario, who says: "I am afraid not only of losing my job, I am also afraid of getting sick, and the question of illness is not an egotistical question, but it looks to a past and a future as well as the present. Anyhow, keep in mind that those who came before us suffered the weight of these heavily polluting companies, and those who came after us will also suffer it."[59]

In both of the above quotations, readers' attention is directed beyond the present moment, otherwise so acute in the text. Whereas Mario's comment extends our collective temporal horizon in multiple directions, acknowledging Taranto's long history of pollution as well as its future impacts, Foschini's statement situates that horizon more precisely through bodies and their functions. Referring to the mammalian transference of milk from mother to baby, he focuses our attention on a sort of transcorporeality thought to be limited to animal interaction, human or otherwise, and often upheld as the most "natural" and indeed healthy of practices. By then underlining the presence of dioxin even in that practice and those bodies, in the milk of mothers and the bellies of newborns, he articulates the transference of pollution and toxicity from one generation to the next. Simultaneously, he again establishes just how profoundly those harmful exterior agents such as dioxin are in fact both "near us and inside of us," to refer once more to Piccinni's text.

Acknowledging that our exterior and interior environments, our skies, rivers, and bodies, are open to the influence of vibrant matter such as dioxin—and that we cannot precisely determine the results—encourages a careful approach to building knowledge. As Alaimo writes, "transcorporeality demands more responsible, less confident epistemologies."[60] It suggests that we need a new vocabulary for an emerging state of being, one based less on singular authoritative narrative and more on the diverse plural narratives prompted by the largely imperceptible but so very present material agents in our midst. By interweaving his own story with those of so many others, Foschini succeeds in presenting a clear portrait of Taranto's toxic reality, a reality that identifies itself in different ways to different subjects.

In electing to tell the stories of these pollutants and these subjects, human and otherwise, Taranto's contemporary storytellers accomplish a great deal. They recognize the ability, agency, and narrative potential of nonhuman matter, just as they give voice to human subjects who may feel they have begun to lose their own agency in the midst of shifting environmental and corporeal realities, as well as an Italian government that does not seem to care. At the same time, they provide readers the opportunity to move from the gathering of knowledge to its sorting through, responsibly and with humility as we must. Once readers come to know Taranto through these texts and others, we can choose to act on our new knowledge. The hope, of course, is that we will speak out against those industrial giants that have caused harm to "those who came before us," in the name of "those who will come after us."

With this exploration of the ways in which narratives can teach readers about ongoing material realities, I do not wish to reduce the creative verbal and visual work of world building via story to an exclusively edifying practice—explaining how Taranto is shaped, who there is ill, and who is to blame. Rather, I wish to underline the ways in which narrative practice and storyworld experience provide crucial space for the imagination necessary to make sense of the complicated realities in which we, or others, live. Storyworlds are particularly crucial when it comes to reminding external audiences that industrially shaped eco-catastrophes affect real living creatures, individual human (and nonhuman) beings with feelings, families, and futures to consider. By offering receptive audiences an experience of narrative imagination, the texts described above not only invite us into deeper knowledge of Taranto's toxicity; they also ask what we might do to combat it.

FIRST PERSON

Angelo Cannata

A few days after first speaking with Daniela Spera in 2016, I met with Angelo Cannata. He had directed me to come find him in Taranto's old clock tower, situated in the top corner of the Città Vecchia's small Piazza Fontana looking out toward the convergence of the region's two bays, Ilva's smokestacks looming in the distance. Featured in Paolo Pisanelli's *Buongiorno Taranto*, the eighteenth-century structure had served in recent years as a meeting space, tourist information center, and home of Le Sciaje, a cultural association Cannata cofounded to document and share Taranto's rich maritime history. Six months after our meeting the tower was closed for a multiyear restoration project, which has since been concluded.

ANGELO: Buongiorno, Taranto! June 30, 2016. My name is Angelo Cannata. I was born in Taranto in 1980, and I'm a cultural guide (*operatore culturale*) in the city. We are sitting now in a historic location, the clock tower in Piazza Fontana, in the part of the city commonly called the Città Vecchia. "Old City" to distinguish it from the new, as the historical-social-urban evolution of the city brought about the realization of an urbanistic phenomenon of decentralization, or rather, abandonment of the city center, and a migration to the periphery of various commercial and economic centers. Because of this, the Città Vecchia has experienced a massive depopulation, harkening back to the first industrial age, followed by the second, which materially abandoned the Città Vecchia.

The "two industrializations," I call them, because those were the periods that really changed the landscape: a landscape that was originally conceived through a strong identification with the city (which was the island) and the Mar Piccolo. Following the two industrializations, first of a military-naval character in the late 1800s, then of a heavy industrial character in the 1960s, the city underwent enormous changes. The first visual, the spatial expansion of the urban fabric, but also of increased population density: a city that had only 30,000 inhabitants until Unification is today a city of 200,000—with a high point in the 1980s of 240,000. It's important to note that the city's urban plan envisions a city of 350,000 inhabitants, because the city was interested in this plan for a strong government-sponsored economy by way of state industry that would have made it one of the most important industrial cities in Italy. And, in fact, it is, because of its strategic military function. Its industry and its [military] port are of international importance.

At the civic level, there's a really important political and cultural consideration to make . . . : industrialization has damaged the landscape, but it's surely made the city more modern and less ignorant. That is, the city has had the opportunity to raise the level of local education. Anyhow, this shows you that it's not just a simple chat that we can have here (*laughing*).

There's so much to say. Keep in mind that we've developed this project here primarily because of two people, a social historian (me) and a biologist who thought to undertake a project about valorizing the history of Taranto's landscape, tied to seafaring, and we developed a project called Le Sciaje. We saw this strong contradiction at the level of citizenry where we worked—prior to the crisis of 2011 and 2012; in "innocent" times, you could say—before the crisis, we thought that there was this cultural urgency. A cultural urgency surrounding a loss of memory of the territory's history, following this major urban evolution, but also a degradation of intergenerational social relations. Older people, our grandparents, our parents—they were almost remorseful when they spoke about the history of our territory. They had even stopped recounting it: they were ashamed of our history. It represented a past of poverty, of shit, basically. And so the factory had made the city better: a vision of quality of life not really oriented toward the future, but instead toward a practical win, and an immediate one. At my age, though, I experienced some alarm [growing up] in the 1990s about the environmental question: there was this environmentalist attitude of looking around in a general way at global warming, but also at the eventual risk of factory-site accidents.

In fact, this morning I was with a group of kids, and they asked me, "But, if the plant explodes, what will happen to us?" And the truth is, we with the biggest factory, with a continuous production cycle, all built up in this zone—we don't have a citywide evacuation plan! If something happens, we should flee the city, right? But they don't tell us that. To avoid causing alarm. Not even after Seveso! Because the Italian legislation doesn't call for a clear emergency plan. In fact, one day there was an accident: total panic. We weren't supposed to open the windows and all of that, and many ran away from the factory because a hurricane had come and it was just a mess. . . . Anyhow, we find ourselves now in this really difficult historic moment in which there's a strong social divide and so . . . welcome!!

Listen, should we go have something to eat? Do you eat shellfish? We'll go to a place where they're "certified," so to speak. Plus it's cheap—and good. We'll have a plate of spaghetti with mussels. *Andiamo.*

[Later, on a drive around the Tamburi neighborhood]

Okay, to better tell the story of the Tamburi neighborhood, you have to also take into account the thing that connects it to the [central] city, the Porta Napoli. Porta Napoli was the northwest entry point. And the city, around the time of the first regulatory plan after the Unification of Italy, needed to expand out toward Tamburi, it had to expand from the northwest. Like every modern city, it expands from the northwest, not the southeast. While Taranto expanded from the southeast and not from the northwest because the Tamburi was originally a well-to-do residential neighborhood. What else reminds us of this? The street names. The names of the streets are all Italian authors. And the name of the neighborhood is really connected to this vision of, how would you say, a development of the arts. It was a residential zone.

Okay so now we're taking a road that from Taranto leads toward the Statte neighborhood. It was once part of Tamburi. This road flanks two important structures, one of which is modern, the pipe factory—do you see it? On the right, though, is the aqueduct, a medieval aqueduct in Roman style. It's the aqueduct that used to bring water to Piazza Fontana, into the city from the Triglie area, underground. All by way of this aqueduct and these historic works of architectural engineering, that are in fact called "tamburi." And so the name of the Tamburi neighborhood is linked, for scientific reasons, we can say, to this passage of water through these arches. But why do the people call it Tamburi? Because this water passed through the tubs and around the neighborhood you could hear the sound of water beating, like a drum. And so the residents are "tamburini," they're drum-people, they're the residents who play the drums. So that's how it is. Except that then after the Second World War, the city of Taranto entered into profound crisis because it had planned its economy of expansion with Unification on an economy of war. During World War I the city grew because the arsenal had work to do, thanks to the economy of war. It kept growing until World War II, after which it entered into crisis. The naval shipyards, which are mostly near the Galeso River but also Tamburi—if there's time later we'll go take a look. So anyway, what happens? Modern industrialization sacrificed this area, because it removed the olive groves, destroyed the farmsteads, and created this: the biggest steelworks in southern Europe. Over here is an old farm where I often take photographers, to explain that in just a few years the city changed its landscape like this, from an agricultural city to an industrial one. Here, hold on, let's stop for a minute.

Do you hear this noise? Twenty-four hours a day there's this noise. It comes from the factory. And so the residents of Tamburi are also stressed-out because of this. You don't hear silence here. Actually, you often hear *Bao! Bao!* Explosions.

You see that blast furnace? Blast furnace number 5. Steel mill and blast furnace number 5. It was built by Nakamura, from Japan, who came to study Italian industry in the final years of state sponsorship. Yes, it was an exemplary model. Why? Because of the integral cycle. Logistically, we're close to the sea, and so they take water from the sea then toss it back out; they take it from the Mar Piccolo, then flush it into the Mar Grande. They use it and don't pay for it. And the steel is of high quality when . . .

[Here I interrupt to ask if children swim in that same water.]

Yes, you've got it. Kids, they have strong immune systems . . . well, some do and some don't.

June 30, 2016

CHAPTER 5

On-screen Engagements
Mapping Plurivocality (and Not)

An unmarked plane lands on a tarmac as an assembled crowd of police and government officials awaits below, brandishing guns and bullhorns. A news reporter runs to join the growing mass, accompanied by his cameraman, who quickly begins to film. Sound is thus far minimal and diegetic, prompting our anticipation to grow alongside that of the characters on-screen. Suddenly, a snazzy synthesizer-driven musical score begins as the plane's door lowers open, revealing a set of stairs. The music swells and speeds, and a middle-aged man in a three-piece suit and bushy facial hair emerges: it is Doctor Otto Hagenbach, director of the state-run nuclear power plant. Hagenbach scans the crowd as the lead official approaches from below and then lurches down to stab the man in the throat. Chaos, naturally, ensues: police and others on the ground rush toward the plane just as a stream of men flow out of it, various handheld weapons drawn in attack. They bear facial lesions that look like extreme chloracne and are dressed in civilian clothing ranging from jeans to business attire. They stab, slash, and otherwise maim anyone who appears in their path, occasionally pausing to drink the blood of a new victim. As Ross Hunter writes, these "antagonists are not technically speaking zombies (or even undead) but rather are the 'infected,' humans that have been turned into bloodthirsty savages due to a radioactive spill."[1] The disfigured men are emblems of toxic embodiment, their actions an aggressive representation of contamination at its most extreme, as they infect others through physical attack.

This scene occurs approximately ten minutes into Umberto Lenzi's 1980 production *Nightmare City* (in Italian, *Incubo sulla città contaminata* [Nightmare of the contaminated city]). As noted in chapter 2, Lenzi's film, which bridges Italy's zombie and postapocalyptic genres, can be read as a reference to the Seveso disaster. Produced just a few short years after the dioxin-filled explosion at the ICMESA chemical factory, *Nightmare City* depicts the aftermath of a semi-urban industrial explosion and subsequent radioactive exposure, including quarantines, animal deaths, and the spread of physical symptoms like the aforementioned chloracne. Myriad reviews and online blog posts claim that Lenzi declared outright his inspiration from Seveso when reworking the script, which originally presented a straightforward zombie tale, to include this eco-angle, but my research has not revealed a first-person quote from the director himself.[2]

Leaning heavily into the gory visuals common in Italian genre cinema of the time, *Nightmare City* explores the mechanisms of fear and confusion that can pervade a community in the absence of clear information. It also acknowledges human culpability in industrial accidents, most notably in a scene I describe below. Calling the film "the most obvious example in Italian cinema" linked to the Seveso disaster, Hunter writes that "its underlying narrative drive is one that explores the potential (albeit extreme) consequences of such a disaster" as "human attempts to defeat possible fossil fuel shortages" run afoul.[3] *Nightmare City* is an entertaining and occasionally rather gross film, one that drips with camp to a postmillennial audience and causes today's university students to both giggle and guffaw, yet its environmentalist warning is consistent throughout. Using what Carter Soles might describe as a "direct and visceral" approach to the structural problem of nuclear power, *Nightmare City* cautions that if we do not respect the more-than-human world, humans will suffer grave consequences.[4]

Toward the end of the film, the aforementioned reporter and his wife, a doctor, find temporary respite in an abandoned roadside bar. Called Miller and Anna, they are played by the Mexican actor Hugo Steiglitz and the Italian actress Laura Trotter. As the couple prepares some instant coffee they've discovered, Anna muses wryly that the substance is "another advantage of human civilization, my dear, like Coca-Cola, or nuclear energy," before adding: "we'd be a whole lot healthier without all that stuff." When Miller wonders whether it might not be better to give up such "advantages," she replies: "Maybe it would be better. However, it is not the fault of science and technology, but of *man*, nature's own human element. We haven't reached the age of the robot yet. This mess could have been avoided." "Yes," he agrees, "but for how long?" Her answer is breezy but direct: "I don't know but one thing is clear: we're partly to blame ourselves. Just think of the life that we have led, up until today: shut up in those ridiculous cities, in a steel and concrete jungle, like machines." Concluding the scene, Miller muses: "all this had to happen to realize the truth," thus alerting viewers to the film's pedagogical mission. Anna's response? An ominous, "Let's hope it isn't too late."

In its condemnation of human thirst for scientific mastery over the more-than-human world (punctuated at film's end by the declaration "nightmare becomes reality," written across the screen before closing credits roll), *Nightmare City* is a film that looks beyond the specificities of Seveso. It considers instead the environmental health risks posed by the types of extreme chemical exposure and nuclear radiation that extractive capitalism facilitates throughout the world. Presumably inspired by a regional event that quickly became national news, *Nightmare City* is in fact a transnational film. Shot on location in Spain as well as studio sets in Italy, it stars a multilingual cast portraying characters with Anglophone names like Miller, Anna, Sheila, and Bob. It was released in an Italian-language dub nearly simultaneously in Italy, Japan, and Germany in December 1980, before a set of English-language releases hit the market three years later. These practices were not uncommon for the era and genres within which Lenzi made his films, as they allowed production crews to work economically yet harness the star power of actors from a broader market, before then releasing their products to the same. More significant for the current study, such practices situate Lenzi's cinematic treatment of contamination anxiety in a nonspecific continental context that might not feel unfamiliar to viewers at large, thus allowing them to imagine something similar occurring in their home communities, just as the narrative remains in the realm of the generically fictional. As such, these practices ensure that *Nightmare City* is not explicitly a "Seveso film," an association that might have otherwise shed glaring light on a community already wary of negative attention.

Most of the contemporary films that I consider in this chapter represent something very different: a practice of making film in and about place, with the express intent to shed light and draw attention. The three titles on which I primarily focus are also, in contrast to a decidedly fictional film like Lenzi's, documentary, a genre that often relies heavily on specificity, as well as a shared assumption of authenticity and truth.[5] A documentary is always *about* something that, viewers presume, already exists in the world—a place, a people, a situation, a story. We must also presume that the creators of any documentary film have made what Bill Nichols calls rhetorical choices, "in which eloquence serves a social as well as aesthetic purpose," that is, choices regarding not only what information is presented (or not) but also how it is presented, and through whose perspective. That said, viewers generally agree that watching a documentary is an opportunity to learn more about the lived world. As Nichols writes, "documentaries lend us the ability to see timely issues *in need of attention,* literally."[6] I do not wish in this chapter to privilege documentary as a more authoritative or informative methodology for telling the stories of environmental crisis. As previous chapters in this book explore, the fictional, the hypothetical, and the speculative are all powerful modes for learning and feeling. Rather, I wish to highlight a series of contemporary films that rely on genre-based assumptions regarding attention to the real, while still exploring unique narrative strategies, so as to draw viewers into deeper awareness of Taranto's true-life toxic tales.

In what follows below, I consider three Taranto-based documentary films shot in and around 2012, the year in which then–Ilva owners Emilio Riva and others were cited for knowingly causing environmental disaster; Ilva's blast furnaces were partially shut down; the first "Save Ilva" decree was issued by the Italian government and full operations thus reinstated; and significant public protest erupted throughout the city, from workers and environmental health activists alike. Presented in chronological order, the films are *In viaggio con Cecilia* (Cecilia Mangini and Mariangela Barbanente, 2014, Traveling with Cecilia); *Buongiorno Taranto* (Paolo Pisanelli, 2014); and *Non perdono* (Grace Zanotto and Roberto Marsella, 2016). Whereas the first two titles are observational-participatory documentaries intercut with archival footage, the third is more experimental, an occasionally poetic documentary held together by a dramatic subthread featuring two faceless figures, to be discussed further below.[7] In spite of these differences, all three films incorporate first-person interviews with area residents. They do so in exploration of what Barca and Leonardi have called "a mental attitude of closure towards the possibility of even imagining economic alternatives to the centrality of the steel plant," and as a means to highlight the work of those in the community who push back against such an attitude.[8] And much like the literary texts addressed in chapter 4, all three films incorporate practices of both mapping and plurivocality, leading viewers through the city and its outskirts and featuring a multitude of Tarantine voices.

Documentary has historically been held in opposition to narrative or "fiction" filmmaking, but these lines are often rather blurry. In the context of Italian film, Luca Caminati and Mauro Sassi remind readers that documentary has walked the boundary between fiction and nonfiction since "the heroic beginnings of Italian cinema," as early progenitors of the medium used props, tools, and emerging technologies to recontextualize historical realities and promote a sense of Italian nationalism. Caminati and Sassi also note the late–fascist era popularity of a genre known as narrative documentary (*documentario narrativo*), a sort of "hybrid fiction" inspired by a series of foreign films, such as Robert Flaherty's *Nanook of the North* (1922), that eventually lead Italian filmmakers toward neorealism. Citing Francesco De Robertis's 1941 *Uomini sul fondo* (*S.O.S. Submarine*) as an example, they write: "the film has a clear trajectory . . . but its many asides enrich the humanity of the story and augment the documentary value of the film."[9] The same can be said, to varying degrees, of the films I address below. All of them are focused on the "humanity of the story," just as they all perform what narratologist James Phelan has described as the rhetorical act of "somebody telling somebody else on some occasion and for some purpose that something happened."[10] It just so happens that multiple somebodies are doing the telling, once again engaging in a practice of dispersive counterhegemonic storytelling.

In sharing Taranto's many and many-voiced stories, the films of Mangini and others contribute to a notable postmillennial resurgence of Italian documentary cinema. Alan O'Leary attributes this surge to the advent of digital technology and its relatively low costs for production and dissemination, which allow people to both film and edit on all sorts of devices, including a handheld cellphone, and share their work through any number of online platforms.[11] Anita Angelone and Clarissa Clò similarly note the relatively low cost of documentary cinema as compared to mainstream narrative film, arguing that "it has the least to lose and often makes the best use of limited resources."[12] In their much-cited introduction to a 2012 special issue of *Studies in Documentary Film*, the authors herald the postmillennial dynamism of Italian documentary cinema and note, in particular, its strong civic and activist stance, which is aided by the aforementioned ease of access to both production and dissemination.[13]

Recent scholarship by Simona Bondavalli reiterates that this has not always been so clearly the case. Her work underscores the expressive freedom of contemporary documentary, by considering instead the close twentieth-century ties between the genre and state-sponsored formulations of national identity produced by the RAI, Italy's national public broadcasting company (just one iteration of an extant relationship between the state and documentary film, as examined by Ruth Ben Ghiat, Marco Bertozzi, and others).[14] Such ties were especially close during Italy's economic boom period, and especially pronounced in documentary programs regarding the nation's environmental and cultural patrimony. Analyzing a series of televisual *inchieste*, multiepisode RAI documentaries, Bondavalli writes that "documentary cinema participated throughout its history in the ambivalent ecologies generated by institutional sponsorships." It did so by celebrating picturesque landscapes and "salvag[ing] rural cultural practices from oblivion" just as it promoted "anthropocentric, often appropriative views of the environment," in programs such as Mario Soldati's *Viaggio nella Valle del Po* (Trip through the Po Valley, 1957).[15]

As I have examined elsewhere, such RAI-sponsored documentary programs, particularly those made in the 1960s, often "equate land and cityscapes to works of art, seeking to elevate the cultural value of the latter through an emphasis on visual aesthetics" that tend to celebrate architectural construction and manipulation of the land.[16] This same time period, encompassing the very years in which the Ilva (then Italsider) steelworks were rapidly being built in Taranto, also saw the production of a significant number of documentary projects celebrating not only construction and production but also extractive industry. Perhaps the most well-known of these is Bernardo Bertolucci's multiepisode televised film *La Via del petrolio* (1967, Oil), once again produced by the RAI but this time in collaboration with Eni, Italy's (multi)national oil and gas company. The three Taranto documentaries that I examine on the following pages share something with their state-sponsored predecessors in that they underscore the deep links between industry and the Italian landscape, alongside ever-shifting notions of national and local identity. Where they differ from those earlier films—in addition to the use of new technologies and alternative platforms for production and dissemination, including not only the internet but also place-based venues such as museums, galleries, and an international festival circuit—is in their critical examination of the ways in which a government-supported industrial giant has failed Taranto.

When ecocritical film studies first emerged around the turn of the current century, much attention was placed on what defined a film as a work of ecocinema. In his seminal essay "Toward an Ecocinema," Scott MacDonald likens ecocinema to reflective nature writing, imagining a text that "offers audiences a depiction of the natural world within a cinematic experience that models patience and mindfulness—qualities of consciousness crucial for a deep appreciation of and an ongoing commitment to the natural environment."[17] As suggested by the words "patience and mindfulness," he emphasizes the pacing of ecocinema, which is often slower than that found in big-budget entertainment films and tends to favor the long take, a type of shot that Nadia Bozak has more recently analyzed as indicative of material access and excess in analogue film.[18] MacDonald also writes of ecocinema's edifying mission to "help to nurture a more environmentally progressive mindset,"[19] much in the same way that Paula Willoquet-Maricondi writes that "ecocinema overtly strives to inspire personal and political action on the part of viewers, stimulating our thinking so as to bring about concrete changes in the choices we make, daily and in the long run, as individuals and as societies, locally and globally."[20]

In key ways, the films that I examine below adhere to this definition of ecocinema: object and text that draws attention to the nonhuman environment (including the damages we have wrought upon it and, by extension, ourselves) and seeks to nurture an environmentalist and indeed activist response in viewers. In approaching them as such, I revisit the econarratological questions posed elsewhere in this study and advocate for the "potential of the [viewing] process to foster an awareness and understanding for different environmental imaginations and experiences."[21] In his work on cognitivist film theory, David Ingram describes a similar approach, one that strongly resonates with the work of James, Lehtimäki, and other scholars of (literary) storyworld cited in previous chapters. Ingram argues that to analyze the ways in which a film might provoke awareness of environmental issues, scholars must consider its cognitive, emotional, and affective reach.

Citing the work of the cognitivist theorist Greg Smith, Ingram explains furthermore that "a narrative film usually works by establishing the viewer's emotional relationship with the protagonist's goals and actions, as well as through lower-level, non-verbal affects, which [Smith] calls 'moods,' produced by stylistic elements such as music, *mise-en-scene,* lighting and color."[22] In their attention to stylistic elements, Ingram and, by proxy, Smith identify the additional tools with which, as compared to literature, film might draw viewers into a receptive experience. Such attention to what are ultimately formal elements in turn recalls the considerations of Donly's eco-narrative or Knickerbocker's ecopoetics, both addressed in this book's introduction, which ask us to consider the ways in which a text's form might reflect the patterns—and pacing—of the more-than-human world. To reiterate a previous argument, the films addressed below are narrative just as they are documentary. While they do not feature singular protagonists, they do provide ample opportunity for viewers to develop an "emotional relationship" with subjects on-screen, whether those subjects are area residents presented through interview, the film's own narrators, or Taranto itself.

Through dynamic practices of both verbal and visual storytelling, each film alternates between traditional talking-head interviews and the particular narrative conceits that mark it as unique: *In viaggio con Cecilia* is held together by an ongoing dialogue between the film's two directors, as well as verbal reflections on and visual citations of Cecilia Mangini's early filmography of the 1960s; *Buongiorno Taranto* is punctuated by the activities of a collectively produced online radio show; and *Non perdono* continually returns to the poetic nonverbal movements of a character whose dress reflects Taranto's religious pageantry traditions, and the active mapping of the Ilva steelworks by his angry alter-ego.

I will start by addressing the first two titles, which were filmed during an overlapping period in 2012. This overlap is made explicit in *In viaggio con Cecilia* by a fleeting shot of Mangini handing Pisanelli a camera at a protest in one of Taranto's main piazzas: she waits center-frame in a crowd, arm outstretched and head turned in anticipation, as he enters from screen-right and smiles briefly at her before glancing directly at us, the film's viewers, then exiting the frame. The two directors' paths had crossed since 2005, when Pisanelli first asked Mangini to be a guest at Cinema del reale, the documentary film festival that he curates in Puglia each year. Among other subsequent collaborations, their codirected film, *Due scatole dimenticate* (Two forgotten boxes) was released in 2020, one year prior to Mangini's passing. A subtle cameo and sweet exchange, the quick handoff in Mangini and Barbanente's film speaks to a shared motivation behind *In viaggio con Cecilia* and *Buongiorno Taranto*. Both films' directors engage with their subjects in a practice of observational filmmaking that is also, to differing degrees, a participatory one, in which "we may see as well as hear the filmmaker act and respond on the spot, in the same historical arena as the film's subjects."[23]

Cecilia Mangini was born in Mola di Bari in 1927, just down the Adriatic coast from Puglia's capital city, Bari, and about seventy kilometers north of Taranto. Although her family moved north when she was a child, Mangini retained a sense of connection to the region and repeatedly returned to film in Puglia in the 1960s, as one of Italy's first and most widely recognized female documentarians.[24] Titles such as *Stendalì—Suonano ancora* (*Stendalì*—They still play, 1960), *Tommaso* (1965), and *Brindisi '65* (1966) explore long-held traditions and changing lifeways throughout the region, in the wake of a newly booming industrial presence. Like so many of the films she directed on her own, from *Ignoti alla città* (1959, Unknown to the city) to *La Briglia del collo* (1974, The bridle of the neck), these short documentaries are particularly focused on class, gender, and labor politics. In this light, Dalila Missero notes that Mangini's 1960s-era work is explicitly inspired by a Gramscian approach to political engagement, which "requires strategies of resistance that start with the observation of everyday dynamics of power."[25] As such, writes Missero, Mangini's films "focus on the everyday lives of the urban proletariat, and on marginalized rural groups, that constituted the forgotten faces of the [economic-boom era] brand-new, wealthy Italy. In this sense her cinema aimed to give visibility to the oppressed and the outcast, and to denounce the controversial aspects of modernization."[26] Missero reads this cinema as both feminist and counterhegemonic, due to its attention to marginalized realities, collective experience, and the practices of everyday life.

The same can be said of *In viaggio con Cecilia*, which revisits similar themes just as it revisits Mangini's earlier films, through interviews, archival clips, and a meandering conversation between Mangini and her codirector, Mariangela Barbanente. The latter was also born in Mola di Bari, although a few decades later, and she shares Mangini's passion for documentary storytelling as both a writer and director of films such as *Sole* (2000, Sun) and the award-winning *Ferrhotel* (2011).[27] Barbanente has explained that she originally approached Mangini with the desire to make a film about the older woman's life and career, but the biographical documentary *Non c'era nessuna signora a quel tavolo* (Davide Barletta and Lorenzo Conte, 2013, There was no *signora* at that table) was already in production, and Mangini thought one film about her was more than enough. Barbanente countered with a new proposal that the two women make a film together, and, after some hesitation, Mangini agreed. As Barbanente shares in *In viaggio con Cecilia*, the original idea for the film was that the two directors would revisit together the locations from Mangini's Puglia-based work, so as to examine the effects of the boom-era industrial dream explored in films like the aforementioned *Tommaso*. While they were location scouting in 2012, the Ilva case broke, and the two directors, both so dedicated to narrative representation of the real, felt compelled to react and record, thus shifting the film's focus to more overtly consider the present day and the complicated relationship between work, environment, and health in Puglia.

In a 2014 interview with the duo, Mangini makes clear that the same interests that drove her earlier work motivated her participation in the film with Barbanente. "I was not convinced about returning to the locations of my films," she says, "but the history of Taranto is the history of Ilva, and before that of Italsider. The meeting between an epoch that seemed to be entering the modern era back when I was filming, and Ilva and the petrochemical industry today, became vital."[28] Barbanente adds: "The microcosm that we recount is for us the point of entry to a broader state of things. There's a message that gets repeated in the film: 'people must react.' But also: 'politics is no longer doing its job.'"[29] Theirs is a film explicitly in, of, and about Taranto, but it is also a film about the failure of Italy's industrial promise for a brighter modern future, about the failure of contemporary political systems to protect the well-being of their citizens, and about the need for community response to and against corporate monsters. The use of archival footage, as well as reflections from Mangini and her various interlocutors about Ilva's early days, makes this the most historically rooted of the three documentaries I address in this chapter, as well as the one most engaged with the question of industrial labor. Environment, health, dioxins, and transcorporeal flows are certainly all present, but anxieties surrounding employment are always at the forefront of the equation.

Like the literary texts discussed in chapter 4, Mangini and Barbanente's film is also a story of mapping. As the title suggests, it is a film about "traveling with Cecilia": a visual diary of Barbanente's trip with Mangini to Taranto and also, briefly, to Brindisi, where the latter shot the early films *Tommaso* and *Brindisi '66*, which both examine the modernizing potential of the Montecatini-Shell petrochemical plant. The women's pace of travel, like so much of their film (its instrumental score, its interstitial shots of land and city-scape), is never hurried but instead almost meditative, serving as a reminder that, as Past writes, "going slow can also be a walking toward, a politics of affirmation, a decision to approach problems by way of physical proximity or cognitive and creative engagement."[30]

In viaggio con Cecilia shows its directors engaged in various modes of proximity, engagement, and travel throughout the film, beginning with an opening sequence in which they drive through the Pugliese countryside toward Taranto. The camera is positioned in the backseat of their car to look out through the space between the two women's shoulders, simultaneously showing them and their point of view so as to position the women as dual subject-storytellers. Later, we see the pair walking: just outside of Ilva grounds, along the waterfront, in the city center, and through the same San Brunone cemetery described in chapter 4. To reach Brindisi, they take a local train: footage of the women in conversation midvoyage is bracketed by shots of Taranto's nearly empty station, a double reminder of travel and its relationship to transition and stasis. One of the film's most poignant sequences, a conversation between Mangini and a struggling mussel fisherman, suggests an additional form of transport, as we see Mangini seated in a gently rocking fishing boat. In his own boat a few feet away, the fisherman, whose livelihood is in peril due to polluted waters, tells her: "We're drowning, we're totally dead . . . people say that they polluted . . . then they say that there's dioxin, no? . . . We haven't understood anything."[31] Mangini laments that she can only hug him *"col cuore"* (with her heart) rather than with her arms, which instead grip the edge of her vessel for balance.

The various modes of travel employed throughout the film confirm its descriptive title, contribute to a particularly active visual charting of local terrain, and underscore the directors' commitment to engaging with a variety of people and places. While they are occasionally alone on-screen, musing and mulling over recently conducted interviews or Mangini's emerging memories, Mangini and Barbanente are most often in conversation with others: men who did or still do work within the factory setting but also additional residents, such as members of the group Cittadini e Lavoratori Liberi e Pensanti; a researcher with Taranto's official tumor registry; the aforementioned mussel fisherman; and young people drinking and socializing in the street on a Saturday night. It is usually Mangini who engages in dialogue on-screen with these various interlocutors, or whose voice is heard posing questions from off-screen when only the interview subject is shown. Underlining the film's attention to environmental health, examples of speaking subjects who appear alone on-screen and thus as privileged subjects include a doctor who speaks of the high incidence in the region of newborns born with tumors, and an artist who describes the rosaries she began making from pills, following her father's death from cancer. It is Barbanente, however, who speaks with the aforementioned researcher, as they stand on a balcony overlooking Ilva's mineral parks; Barbanente who approaches workers outside of the petrochemical plant in Brindisi, trying to find one willing to speak on camera; and Barbanente who often walks beside Mangini, or perhaps just a step behind, in quiet observation.

As in the other two Taranto documentaries addressed below, the interview sequences outlined above allow *In viaggio con Cecilia* to feature the voices of a vast array of residents in both Taranto and Brindisi. Largely in contrast to the other two titles, however, the directors of *In viaggio con Cecilia* enter overtly into the conversation, their voices and bodies often present but never dominant. This steady presence conveys the directors' authority, or better yet their authorship, but it also serves as an invitation, to both on-screen interlocutors and viewers alike.[32] By positioning their bodies on-screen in many of the film's interviews, as well as in scenes dedicated to their own ongoing dialogue, Barbanente and Mangini serve as the primary visual thread between each stop on their journey. This allows them to occupy a delicate liminal position. On one hand, they directly engage with place-based community in demonstration of what Valeria Castelli calls "an ethical care for the social collectivity." As Castelli explains, by filming themselves in the place they seek to represent, and specifically by showing their bodies on the ground alongside those of their cosubjects, the directors act as "part of a polis . . . engaged with collective social and political issues."[33] What's more, in a visual reminder of horizontal solidarity, they are almost always shot parallel to these interlocutors.

On the other hand, they cannot help but to travel through Taranto and Brindisi with the remove that inevitably accompanies the role of documentarian, observing and analyzing at the same time that they interact. Rather than detracting from the film, this is a crucial element of *In viaggio con Cecilia*'s efficacy as teaching tool. While their physical grounding allows the directors to act as "part of the polis," their actual distance from that polis as critical observers allows for viewers, who may not be part of the community represented on-screen, to align ourselves with the directors, and with Barbanente in particular. As she and Mangini travel through the film and its locations, their questions reveal them to be informed outsiders seeking to understand the community's current ethos. Following along, and privy to their private reflections, we begin to place ourselves in their shoes: neither fully outside nor inside the community group, and thus necessarily engaged and attentive observers. As the film progresses and we enter further into its storyworld, we realize that we too are *in viaggio* with Mangini, just as Barbanente is, and we begin to cultivate our own sense of care for the places and people with whom the women interact.

It is worthwhile to underline that while the film privileges Mangini's reflections in light of her particular connection to Taranto and Brindisi's industrial histories, it is Barbanente who serves as limited narrator, directly addressing viewers in voice-over at the film's beginning and end. Castelli notes that these carefully placed instances of voice-over help to establish Barbanente's cinematic authority and role as codirector of the film. At the same time, she writes, their content indicates to viewers "that Mangini's point of view is worthy of the viewers' trust."[34] Occurring approximately seven minutes into the film, Barbanente's first direct address to viewers explains the film's setting in terms of both time and place: "It's summer, and a judge in Taranto has ordered the sequester of the largest steelworks in Italy and the arrest of the proprietor, Emilio Riva, for environmental disaster. The jobs of many workers are at risk."[35] She then explains the connection between herself and Mangini, both filmmakers who care deeply for their native Puglia, before stating the following about her collaborator: "I want to try to understand what's happening from her point of view."[36] Regarding fictional literature, Marco Caracciolo notes that "in first person narrative the narrating character's perspective will tend to become salient in readers' experience of the text."[37] The same principle applies here. Although Barbanente does not narrate through the majority of the film, the fact that she addresses viewers directly at its start (and then again at its end) shapes our receptive experience, and thus our shared inclination to also try to see from Mangini's perspective.

Our affinity with Barbanente and her Mangini-focused gaze is first cultivated visually in the film's opening minutes by some very deliberate framing. After a series of film-initial tracking shots featuring wind turbines nestled throughout a rolling golden-brown hillside, the camera cuts to look out from the backseat of a car. The directors occupy its front seats, Barbanente behind the wheel, and speak about the changing Pugliese countryside: they agree that the recently added turbines make it look dynamic, more alive than before. The camera is positioned to share their view, the same landscape from seconds earlier, while also framing the back of the women's heads and shoulders, dark against the scene beyond their windshield. There is also another corporeal element present on-screen: Barbanente's eyes, reflected center-frame in the car's rearview mirror as they dart between the road in front of her and the diminutive Mangini to her right. Engaged as she is in dialogue, Barbanente never glances directly in the mirror and so we never have the sense of locking eyes with her.

Nonetheless, the scene's geometric framing visually compels us to connect with her and her gaze, which is equally directed to the terrain before her and the interlocutor beside her.

As Mangini herself readily states throughout the film, her view of Taranto and Brindisi in 2012 is deeply colored by her observations of those same places and their people in the 1960s. At that time the Ilva (then Italsider) and Montecatini-Shell plants were first coming into being: the promise of a brighter future for Italy's South through modern technology was the dominant ideology promoted from on high, and area residents were transitioning away from agricultural lifeways and into factory work, reaching toward that promised future. In the films listed above, as well as in *Essere donne* (1965, Women) and others, Mangini documents this boom-era transition while critiquing what Missero describes as "the actual structures of power and oppression that affect a subaltern group,"[38] and what Mangini herself calls the women's "hopeless situation" due to "the production line, the compartmentalization, the short timescales, [and] the confirmation of Gramsci's teachings on Fordism," which suggest that the collective promise of routinized labor cannot succeed in a capitalist society.[39]

An example of such critique can be found in the first scene of a segment dedicated to southern Italy in *Essere donne*, which examines working conditions for women throughout the country. As I discuss further below, this is the first archival clip used in *In viaggio con Cecilia*, just after the aforementioned opening sequence. As the scene begins, we see a group of women filmed at ground level in relative close-up, harvesting olives. A narrator (he is male—the gender of narrative authority in the 1960s) explains in voice-over: "Today big industry looks out even from among the olive groves of the *mezzogiorno*, but the industrial development of the south is slow, insufficient. The olive harvesters know that factory work is a necessary passage to avoid backward working conditions, as well as a patriarchal relationship with their families and with men."[40] The camera then zooms up from a woman hunched in the dirt gathering freshly fallen olives to an industrial tower in the background, a barbed wire fence clearly visible between it and her. With this the narrator concludes, "But the factories are few. This is why in the south even the women are migrating," and the camera cuts away.[41]

In this brief sequence, verbal narration establishes the following: industrial manufacturing has invaded the southern landscape, and manual laborers feel they must enter the factory in order to keep up with modern society, but southern industry does not meet the needs of those laborers; it is insufficient. The film's visual effects complement the narration by underscoring the ways in which industrial manufacturing looms also as a physical threat to both a naturally abundant nonhuman nature and the humans in its midst. At the same time, this sequence and indeed the whole film highlight the shifting bounds of what is considered women's work in boom-era Italy, as emphasized by the narrator's use of the word "even" in the line quoted above. That this is the first archival clip used in *In viaggio con Cecilia* informs viewers that Mangini and Barbanente are still attuned to questions of gender related to the factory setting, which is still so visually male. By highlighting working women early in their film through the clip (women working in agriculture but destined to move into industry), they provide a counterpoint to the present-day interviews with male factory workers in later sequences, reminding viewers that women, too, perform various types of labor in a place like Taranto—industrial, affective, and beyond. This subtle but immediate nod toward a feminist sensibility early in the film is emblematic of much of the directors' narrative approach in *In viaggio con Cecilia,* which is largely one of showing rather than telling.

This approach extends to the film's environmentalism. While the directors verbally share their opinions only in select moments (that type of expression is mostly reserved for their interlocutors), they readily use the formal elements afforded by their medium to direct viewers' attention and affective response: cutting, framing, focal distance, score, and so on—the "lower-level, non-verbal affects" that Greg Smith calls "moods."[42] Consider the following. As described above, the film begins with a series of lovely landscape shots before introducing Mangini and Barbanente, who are driving through that same landscape. Following their drive, and before the clip from *Essere donne,* the filmmakers disembark at the ancient Roman bridge over the Ofanto River, northwest of Bari. When Barbanente asks Mangini why she wants to start the film there, the older woman explains that the place reminds her of "the Puglia of her youth." The camera then cuts to a shot of bright-green grasses subtly vibrating in the river as it flows. After a quick jump back to the women in profile, leaning against the bridge's edge, the camera offers another view of the river in close-up: small rapids pouring over a series of rocks, all various light-dappled shades of brown, the leaves of a tree framing the left side of the screen. Once more, the camera cuts to the women, now filmed in a long shot as tiny specks in the top third of the screen, the stone bridge that supports them stretching from left to right, a moody sky above and the river running below. A few seconds later, the aforementioned archival clip rolls, ending on a somber note about transition from field to factory. It is immediately after this clip that we hear Barbanente's voice-over, in which she speaks of environmental disaster.

This sequencing explicitly establishes *In viaggio con Cecilia* as a work of ecocinema. What Scott MacDonald has written of duration in Peter Hutton's landscape films can be said for shot order here: it "is a way of arguing for the comparative *importance* of what we're seeing, and of the manner in which we're seeing it."[43] By visually introducing the Pugliese landscape before they show themselves, and by filming it in clear focus while they are instead shot over the shoulder, in profile, or at a great distance, Mangini and Barbanente posit Puglia's more-than-human nature as a primary subject of their film. It is *important* that viewers clearly see the rapidly flowing Ofanto River and the region's rolling hills, dynamic and changing with the addition of a renewable clean energy source, so that we understand just how precious the landscape is. When this establishing sequence then concludes with a fifty-year old voice-over warning of a transition from field to factory, followed by a modern-day confirmation of environmental disaster, the message is clear: this is a land worth preserving, and factory culture has failed it.

A similar message permeates Paolo Pisanelli's *Buongiorno Taranto* (2014), delivered by a younger and decidedly more collective voice. As noted above, Pisanelli's filming schedule overlapped with Mangini and Barbanente's, although he actually began shooting even prior to the events of 2012, when the Taranto-based activist group Cittadini e Lavoratori Liberi e Pensanti organized their first large assembly in August 2011. Like Mangini and Barbanente, Pisanelli is both cinematographer and director, and had made numerous documentaries prior to *Buongiorno Taranto*, including a 2006 study of the Pugliese tradition of pizzica (tarantella) music and dance and a 2010 film on the aftermath of the earthquake that had devastated the city of L'Aquila one year prior.[44] Again like Mangini and Barbanente, Pisanelli is Pugliese by birth but not from Taranto; he hails instead from Lecce. As he explains in a 2012 interview, this means that for him Taranto had always been "an unknown but fascinating realty."[45] His positionality, proximate but not immersed, allowed him to see Taranto as a representation of Italy at large. He states: "Taranto, like L'Aquila, is a mirror of Italy: the result of the worst political decisions possible. It's a city that is paying the consequences of a model of industrial development completely in conflict with health and the natural environment, that everyone supported for as long as they could.... I was struck by the *mea culpa* of the citizens, and even the workers, who admitted to having averted their gaze for too long."[46] Pisanelli's use of the past perfect tense in this final sentence is crucial, as it suggests that many in the community now recognize their challenging situation and, hopefully, are trying to change it. As he explains in that same interview, and as many subjects in *Buongiorno Taranto* reiterate: "The situation seems desperate, but Taranto doesn't only have a lost beauty. There is a beauty that is still present, and worth being valorized."[47]

Like Mangini and Barbanente's film, *Buongiorno Taranto* shows a city torn, angry with the industrial giant that has sickened its waters, woods, sheep, and humans, yet heavily dependent on that same giant for so many livelihoods and unsure of what might come next. It also shows a place that is visually stunning in so many ways, from its bright-blue water (a recurring visual motif) to its evocative Città Vecchia, and a people who are as enthusiastic and innovative as they are frustrated. The film is marked by color-rich exposures, smooth rapid cuts, and a vibrant musical score; a high-speed cover of the Nancy Sinatra classic "These Boots Are Made for Walkin'" serves as suggestive power anthem. It is also marked by a steadily paced rotation between scenes featuring the making of a web-based radio program; interviews with area residents; handheld protest footage; panning views of land and city scape; and, sparingly, archival footage of midcentury Taranto, from films like the Italsider-produced *Acciaio fra gli ulivi* (Giovanni Paolucci, 1962, Steel among the olives). The film's beginning and end are also marked by ambient images of thick, smoky clouds streaked with red, no doubt a visual reference to steel production but also eerily suggestive of nuclear explosion, or perhaps of the "civil war" that one older interview subject declares is imminent.

The title *Buongiorno Taranto* will surely ring a bell for viewers of a certain age, as it recalls the 1987 American film *Good Morning, Vietnam*, directed by Barry Levinson. A comedy-drama starring actor Robin Williams and set in 1965 Saigon, Levinson's film is very different from Pisanelli's in obvious ways, and yet it offers a suggestive point of reference. Williams's character, based on the real-life Adrian Cronauer, is a disc jockey for the United States Armed Forces Radio Service. He strives to keep listeners' spirits high with great humor and energy but is ultimately compelled to speak out on-air against what he sees as the true horrors of the Vietnam War, thus defying the official U.S. narrative. If readers have never seen Levinson's film, it is still quite possible they have heard references to, or indeed imitations of,

Williams's now iconic drawn-out greeting (it is a yell, really) for which the film is titled. This is enough to understand the pathos that it conveys; a spirited cry brimming with action, alongside a palpable trace of irony—for can a morning ever be good in a war zone? And what if that war zone has occupied an already-colonized land?

The on-air voices in *Buongiorno Taranto*, hailing from a seemingly fluid collective rather than a single individual, convey something similar. Their daily greeting, much like that of Cronauer/Williams, also serves as a literal wake-up call for their fellow citizens, what Ivelise Perniola also describes as "a heavy action of counter-information in the face of a passive population, historically habituated to silence and acceptance."[48] Theirs is a wake-up call to acknowledge not only the current day but also the reality in which they live, caught between workers' strikes and *"diossina, Mercurio, IPA, Berillio, Benzo-Apirene, PCB, Amianto, Piombo, Arsenico, Radioazioni nucleari,"* as listed by one on-air voice five minutes into the film. Not quite DJs, the Tarantini who run *Buongiorno Taranto*'s radio broadcast are program hosts (*conduttori*): they report on local happenings, interview residents, and hold discussions in their studio space, situated in the sixteenth-century clock tower of Taranto's Piazza Fontana, gateway to the Città Vecchia. While viewers see quick shots throughout the film of the various hosts speaking from behind a microphone or walking city streets, we also hear their voices tracked over observational footage of Taranto and its residents. In this way, their audio functions simultaneously as narrative voice-over for viewers and diegetic soundtrack for subjects on-screen, who may be listening to the program as they go about their day.

Like Williams's character in the Levinson film, the *Buongiorno Taranto* radio hosts speak with an infectious energy as they try to reach their fellow citizens; and like *Quindici passi*'s Giuliano Foschini, they are explicit about the value they place on story as means to knowledge. As one member of the collective begins her broadcast to listeners: "Imagine the sea, a window, a table, a computer, a microphone, and many stories to recount. Good morning *Taranto nuestra* [our Taranto]. Are you awake?? Wake up, Tarantini!"[49] Not only does she repeat what is by now a familiar refrain—Taranto's true stories are many—but she draws listeners into a state of engaged reception by asking them to use their imaginations, conjuring a scene that they cannot actually see. While this is an effective way to set the scene for listeners, it also serves as a reminder to the film's viewers, those of us who *can* see the speaker in the precise setting she has described, that something can be very real even when not presently visible—whether this something refers to the collective energy of concerned citizens, or the circulating toxins listed at the film's outset.

The members of the radio collective function in Pisanelli's film much as the chorus might in a Greek tragedy, helping to narrate the action but, more significantly, to offer commentary and opinion. There are perhaps eight primary members, women and men, most of them young adults. As opposed to a traditional Greek chorus, we do see their individual faces: one is the aforementioned Angelo Cannata, who has confirmed in our correspondence that the others are, like him, local residents active in community movements at the time of filming.[50] While they might thus be recognizable to a Tarantine audience, they are never identified by name, and their voices are often cut together, audibly distinct but flowing from one to the next.

Paolo Pisanelli can also be spotted on-screen as part of this group. He too is never identified by name, just as he never addresses viewers via voice-over or gaze, save one playful glance in the film's final minutes. Instead, he appears as an advisor to the radio hosts, engaged in collaborative dialogue in the group's clock tower studio space. He first appears about ten minutes into the film, working with the group to flesh out the concept of the citizen-driven radio program. He is accompanied by a sound technician and camera operator, who hover gently in the background, their filmmaking tools only a small part of a mise-en-scène otherwise focused around speaking subjects. He next appears for just a few seconds toward film's end, again surrounded by the radio hosts but now behind the microphone himself as he introduces the first official episode of their collective radio program. Much in the way that Barbanente's voice-overs bracket *In viaggio con Cecilia*, the scenes of Pisanelli (and his crew) bracket *Buongiorno Taranto*, subtly confirming his authority as filmmaker at the same time that they show him to be "part of the polis," his body in frame just as it is on the ground. Furthermore, by overtly showing the filmmaking process in the first, longer appearance, *Buongiorno Taranto* dips into the tradition of the "making of" documentary, thus confirming its status as a self-conscious story about stories, or rather a story about the process of story itself.

In the first scene mentioned above, the group is discussing a common need in Taranto for open dialogue as a sense-making practice, and the accessibility of radio as a means of dissemination. After one member explains: "There's a need for awareness. How do you spread awareness? Only through communication," another emphasizes: "What's great about the radio is that it's free and can reach everyone." Pisanelli, who is otherwise silent, chimes in to add: "The person we meet who has something to say, *that* person could [also] be a host of 'Buongiorno Taranto,' of this web radio." Their common interest in widespread communication, as well as the belief that anyone can and indeed should tell their own story, is repeatedly echoed by various hosts throughout the film. As one says to listeners on air midway through: "Open your eyes! Take action! Turn on your technology, your mobile phones, tablets, photo cameras, video cameras—and recount your territory!"

Pisanelli's suggestion to the group that anyone might become a host of the show and thus participate in its collective authorship underlines his interest in a community-driven plurivocal narrative. At the same time, that subtle exercise of his role as director gestures toward a fact that remains otherwise unarticulated on-screen: the radio program was actually and exclusively created for the sake of the film. While the hosts' monologues are spontaneous and unscripted, as are their conversations with each other and their interviews with area residents, the web radio show *Buongiorno Taranto* is a framing device conceived by Pisanelli. The director has explicitly attributed this choice to what he saw as a need for a more pronounced plurivocality in recounting Taranto's stories. In a first-person essay for CON magazine.it (of the organization CON IL SUD, [With the South]), he explains:

> The knot to untie was always that existential dilemma: health or work? The complicated relationship between factory and city requires an in-depth analysis from many points of view: industry, economy, pollution, future prospects. And so I decided to recount the *here and now* of those who live in the city, and to enact a narrative mechanism that expressed itself through the words and thoughts of the Buongiorno Taranto hosts: a nomadic web radio, dynamic and unpredictable. A really original radio show that invited every person off the street with something to say to be one of its hosts. An unbounded approach, free from the frameworks that, in other modalities, would turn protagonists simply into people to "be interviewed."[51]

In that same essay Pisanelli writes: "For me the radio is a magic box, an invisible ray that spreads through the air, a sort of illumination that allows you to build relationships between strangers—all you have to do is gather around a microphone and record thoughts and words." In the context of the present study, his description of radio's dissemination—an invisible process of spreading through the air—reads as strikingly similar to the spread of dioxins and other harmful pollutants emitted by Taranto's Ilva steelworks every day. That he punctuates the description with a final image of a group gathered around one single device to share thoughts and words speaks once more to a belief in the power of story and community to combat such pollutants, while acknowledging their patterns and flows.

More than the other two documentaries I analyze here, *Buongiorno Taranto* offers a truly plural narrative voice in that apparent authorship is almost totally decentralized, with the exception of Pisanelli's brief and subtle appearances on-screen. In this, it is also a more optimistic film than the other two, as it models a path forward that is dynamic and collaborative, like that explored by a number of local groups featured toward film's end, such as the *officine tarantine* cooperative community center. That Pisanelli fabricated the radio program to serve as the film's framing device recalls the malleability that has always been present in documentary film practices. It also underscores the fact that, as one of the radio hosts expresses in the brainstorming session described above, "it might just be time for everyone to say what they want." Rather than serve as critique of the community (a suggestion, perhaps, that an outsider had to come in and create this forum, as they had not yet done so themselves) the *Buongiorno Taranto* web radio elevates a spirit and initiative toward collective communication and representation already present in Taranto.

One of the most charming groups of speakers handed the microphone in Pisanelli's film as it nears its end is a collection of local boys who appear to be between eight and twelve years of age. As the camera cuts from ambient footage of neighborhood housing to the group of six shirtless friends seated by the water's edge, the segment's host (one of the six) holds a microphone in the middle of the giggling group. He begins: "Buongiorno Taranto! Speaking to you is the . . . the . . . Città Vecchia crew! And the subject we are speaking to you about is the sea." Turning to the boy on his right, he asks in dialect: "What do you think of the sea?" The boy declares that it is beautiful, and that every time they swim in the sea they have a new experience. Then, turning to his left, the host asks another boy, "And the pollution?" The reply: "There's tons!" Finally, he asks the smallest of the group, "What job do you want to do when you grow up?" With some help from his friends, the boy replies that he wants to be a fisherman (or rather: he wants to "throw" himself "in the water and grab fish"). The group expresses their approval with laughter and jovial jostling, before their fearless host declares that he will now take whoever is behind the camera swimming.

Suddenly, the camera plunges into the water along with the boys and the scene ends with a series of gorgeous underwater shots. An accordion score plays as members of the "Città Vecchia crew" flit about in the blue, nimble as dolphins in what appears to be their natural element. Their appreciation of the sea, their recognition of its polluted state, their desire to still toss their bodies into it with glee and mutual dependence years into the future all merge, as these shots devoid of dialogue offer viewers an ecocinematic moment of visual delight. At the same time, they serve as recollection of their city's mythic founder, Taras, and as plea to preserve its contemporary environment.

The final Taranto documentary I consider in this chapter guides viewers through the city in 2014, two years after the events featured in the films addressed above. *Non perdono,* by the artists Grace Zanotto and Roberto Marsella, features a patchwork of interwoven threads based in speculative discourse, representative physical movement, and traditional realist documentary, making an argument for the ways in which a less-than-realist approach can easily "provide interpretive schemes for the world."[52] The film features the poet, actor, and activist Luigi Pignatelli in the interpretive title role, former head of Ilva railway operations Daniele Amati as a single-minded narrator, and a host of local Tarantini in first-person interviews. Offering a complex portrait of a community seeking to understand what comes next, *Non perdono* asks what has since become a familiar question, What might Taranto be, in both the present and the future, if Ilva were to no longer exist?

A completely independent, self-produced film, *Non perdono* is rougher than the other two documentaries considered here—its sound uneven, its transitions unpolished. As Zanotto told me: "We used our own equipment and resources, because for us it was important to do it and we did not have time or patience to search for funding. The project was born from a conversation we had at an art opening: the idea at the heart of it was that that factory of suffering and pollution should be stopped, and our anger was strong enough to make it explode."[53] It is that same forceful energy that fuels the film and that, along with its merger of modes realist and otherwise, makes it an important object of study for understanding Taranto's many tales.

As suggested above, this film, like the other two already considered, falls squarely into the realm of the plurivocal. In addition to the voice of Daniele Amati, viewers encounter the voices (and faces) of many other Tarantini through spoken interview. Subjects include Daniela Spera and Vincenzo

Fornaro, son of the aforementioned Angelo Fornaro, as well as the hip-hop musician Fido Guido; the president of World Wildlife Federation (WWF) Taranto; a loquacious wig shop owner; a cemetery florist; a few older residents skilled in traditional craftwork like painting and restoration; a set of former Ilva employees; and a host of young people engaged in resuscitating the city, especially the Città Vecchia and Tamburi neighborhoods, through art and creative expression. As is also true in texts explored in previous chapters, the cumulative effect of these subjects' testimonies suggests that some experiences in Taranto are shared (almost all subjects speak of health or environmental crises) while other experiences and opinions are unique (a young ceramicist speaks of the beauty and excitement in collective urban renewal, while the florist describes Tarantine existence as hopeless and the protests around him as useless).

Non perdono also participates, again like the other texts discussed here and in chapter 4, in a visual mapping of the city. This mapping occurs naturally, if not also disjointedly, as the camera cuts from one interview location to the next, whether inside an artist's studio or perched under a tree on the Fornaro family's land, with Ilva's smokestacks off in the distance beyond a speaking subject's shoulder. By constantly shifting locations, the interview scenes allow viewers to visually piece together an image of Taranto including diverse neighborhoods, a massive industrial area, and striking land- and sea-scapes. This is not unlike the sort of mapping that occurs in *Buongiorno Taranto*, one that features multiple sites throughout the city but does not clearly explain how to get from one to the next. It is the sort of disjointed visual presentation of place that might confirm Certeau's description of places as "fragmentary and inward-turning histories, pasts that others are not allowed to read, accumulated times that can be unfolded but like stories in reverse, remaining in an enigmatic state."[54] Not unlike Piccini and Argentina's novels, discussed in chapter 4, it reminds viewers from outside the community that we must remain active and engaged in our receptive practice for the story to make sense.

There is another sort of mapping in *Non perdono*, however, one that offers clearer coordinates and is directly connected to the film's aforementioned move beyond strict realism, by using the hypothetical and the gestural to complement the realist. This mapping is conducted by two oddly twinned figures who work together to offer the film's frame narration: Daniele Amati, who is identified as having at one time directed Ilva's railway operations, and the aforementioned white-cloaked figure portrayed by Luigi

Pignatelli. Whereas Amati maps verbally and manually (described further below, he literally draws a map throughout the course of the film), Pignatelli's character does so by way of silent physical movement, sweeping through Taranto as both an embodiment of potential salvation and a reminder of transcorporeal flows. Each of these frame figures is introduced at the start of *Non perdono*, just after the film's opening credits.

Viewers see Pignatelli first, in a moody over-the-shoulder shot along the water's edge, the pink evening sky fading to black, before the camera quickly cuts to a new shot of him, now in a whitewashed interior courtyard. Dressed in white robe and full hood, face obscured, he is shown here in a frontal medium shot centered on a small red ball held at waist height in his gloved hand. His clothing, including a rimmed black hat, identifies him as a *perdono*. *Perdoni*, or *perdune* in local dialect, are figures from Tarantine religious pageantry that feature prominently in the events of the Easter Holy Week. Pilgrims from among the Confratelli di Carmine (Brothers of the Church of Carmine), they appear in the early afternoon of Holy Thursday, walking barefoot and cloaked between the many churches and small altars and memorials to the dead spread throughout the city. As they do so, they trace out a spatial trajectory that is as unique to Taranto's more-than-human geography as it is embedded in its cultural traditions.

The appearance of the *perdoni* is the first act of the Holy Week that involves the whole community, as residents come out onto the street to bear witness and offer greeting, and it is said to symbolize the walk of pilgrims who once traveled to Rome in search of God's pardon. Notably, the *perdoni* are always paired, appearing in twosomes referred to as *poste*. Many sources cite the distinction between the *posta di campagna*, the pair of perdoni who walk the outer parts of Taranto that once were open countryside, and the *posta di città*, the pair that remain in the Old City center. As Linda Safran writes, each pair is "greeted in the street with *u salaamelecche*, a respectful inclination that surely derives from the Arabic *salaam aleikum* or the Hebrew *shalom aleichem*, peace be with you."[55] As they trace out a path each spring of pardon desired and peace received, the *perdoni* reaffirm Taranto's routes and byways through their walking ritual, just as they take on the city's collective sins.

Pignatelli's *perdono* in the film does something similar. In between interview segments, viewers see shots of this figure in various locations, almost all of which are exterior and allow us to see recognizable Tarantine landscapes in the background, even if we don't have deep familiarity with the city. This might mean that he is walking on sand along the edge of the Mar Grande, with the Città Vecchia behind him, or that he is walking through fields, perhaps of the Fornaro family's Masseria Carmine, with Ilva's blast chimneys in the distance. In one shot we see his bare feet descending a set of stone steps; in others he floats past small white altars, again along the water's edge; in yet another he silently bangs his fist against one of the many death notices plastered on the outside of a rundown-looking civic building, parked cars edging in from frame-right. Most frequently, however, he is framed against Taranto's more-than-human landscape, whether aquatic or plant-rich, a formal choice that links him to the tradition of the *posta di campagna*. In one particularly evocative scene, he is shot at close range in an olive grove, the camera's focus blurred as his gloved hand caresses the leaves of a tree that may well be affected by the *Xylella fastidiosa* bacterium that has plagued the Salento region since at least 2013. Here, as in other scenes, the *perdono* pauses to place his hands over his face in a gesture of grief. He moves slowly and sensuously and utters no sound, his scenes tracked over either with audio from whatever interview preceded them or a meditative piano score.

As described above, the word *perdono* is a synonym for "pilgrim" in the context of Taranto's religious traditions. It is also, however, a common noun in Italian meaning "pardon," "forgiveness," "absolution," or "relief." It can be used as a formal supplication, as in, "I beg your pardon," or the first-person singular conjugation of the verb *perdonare*, "I pardon." The title of the film *Non perdono* can thus be interpreted as a declaration ("I do not forgive"), a negation ("I do not beg your pardon" or simply "Non-forgiveness"), or an identification of Pignatelli's white-cloaked character as an alternative figuration of traditional lifeways (an *anti*pilgrim). I like to imagine that it encapsulates all of these meanings and more. After all, in his dynamically gestural presence, Pignatelli's (non)*perdono* exists beyond the bounds of definitive language. He serves as a purely visual guide for viewers, leading us through Taranto as he underscores the centrality of the more-than-human environment to the film's storyworld (what Adrian Ivakhiv might instead call *film*-world, "the available worldness of a given film") and, in turn, to the crisis unfolding in real-life Taranto.[56]

Martin Lefebvre and others have written on the ways in which landscape often exists outside of narrative in a film, offering viewers a space to pause their receptive experience.[57] In *Non perdono* this is decidedly not the case: the *perdono*'s constant presence in the landscape, combined with his directed gaze and his refusal to remain still, again pushes us to remain active observers at all times, ever-focused on the story that is ever-unfolding. In his seeming omnipresence throughout the Tarantine landscape, the wide expanse of his arms, his probing touch on all sorts of living matter, he charts out the same course as the dioxin addressed in every one of the film's interviews: seemingly everywhere at once and yet impossible to pin down. In this he also serves as a visual representation of a transcorporeal ontology that, as Alaimo writes, "contests the master subject of Western humanist individualism, who imagines himself as transcendent, disembodied, and removed from the world he surveys."[58] He is so fully of the material world on-screen, moving through it as it moves through him, that it's no wonder he continually expresses grief.

Shortly after viewers encounter this singular *perdono* figure in the first minutes of Zanotto and Marsella's film, we see Daniele Amati's character. As is also true for Pignatelli, we do not see Amati's face. This is not because it is obscured by cloth but because he is only ever shot from above, leaning over a writing desk as he draws out a map of Ilva territory. Although we hear his voice, viewers only see his hands and bare forearms, as he draws and labels coordinates with a thick black marker in chronologically sequential shots intercut throughout the film. His canvas, a plain sheet of paper, is held down by a pistol on one side, and a row of cylindrical black canisters on the other, fuses sticking out as though they might be hand-held bombs. A detailed commercial map of Taranto is tucked under the pistol, generally positioned in the top left of the screen, while a miniature figurine of a *perdono* sits just below the row of canisters, barely visible at the very bottom of screen-right. As Amati draws, he describes the steelwork's visual layout—its dimensions, its points of entry, and its most potentially explosive areas—eventually explaining the most effective way to blow it all up.

About fifteen minutes into the hour-long film, Amati and Pignatelli's characters become visually linked. Following an interview segment with Vincenzo Fornaro, in which he recounts his family's battles with both Ilva and cancer, the camera cuts briefly back to Amati, or rather to his hands, as he continues his mapping project. Ten seconds later it cuts to a new frontal angle of someone seated at the same marble table in whitewashed courtyard, drawing with black marker on white paper. He is dressed in the *perdono*'s white robe, gloves, and hood, but missing the additional accessories such as hat, cloth belt, and cords that we have seen on Pignatelli's character until now. A distracted viewer might think that we have finally been shown a fuller view of Amati at work, now in simplified pilgrim's dress, or that Amati and the *perdono* are in fact one and the same. Upon closer look, however, we see that the image being drawn is a simple tableau of abstract lines—something that gestures toward a map but lacks any detail. The repeated cross-cuts that follow throughout the rest of the film, showing first Amati's clear mapping and then Pignatelli's abstraction, connect the two figures as dual narrators, one verbal, the other not. In their pairing, they nod toward Taranto's Holy Week *poste,* the two *perdoni* who walk the territory together, just as they provide the film with a poetic frame tale, Amati voicing the enraged spirit of Taranto's present day, its nonforgiveness, while Pignatelli pulls upon its rich history to grieve for what has been lost. Together, their unique acts of mapping, of pilgrimage, allow viewers to look toward a potential future.

In the film's final minute, Amati's composition complete, he places the *perdono* figurine right at the water's edge on his map, in a NATO-controlled location along the border of Ilva territory where nuclear submarines are rumored to rest. With the four explosive canisters placed as markers at other locations throughout the map, the setup is now complete: "And here we are: Capo San Vito, cargo port, Ilva, Oxygen, Eni. LADIES AND GENTLEMEN: WE HAVE SHAKEN [blown up] TARANTO!" In both his language and his swift strokes on the page, Amati is as decisive as Pignatelli is meditative—drawing out a map not so that bodies may walk the terrain but so that they might destroy the "monster" that has been poisoning it. What he describes is a violent sort of eco-terrorism but also a potentially liberatory act, defiantly reclaiming agency through the removal of a massive source of danger to the nonhuman environment and the bodies in its midst. It is an act that says, "I do not forgive," just as it simultaneously offers *perdono,* relief.

By first visually plotting out Ilva's grounds, then eventually offering directions for its destruction, these framing sequences transition from the same sort of reality-based spatial mapping discussed in the previous chapter to a register that is instead speculative and, we assume, fictional. Like the exterior shots of Pignatelli's *perdono* in the Tarantine landscape, this related thread pushes the borders of traditional documentary while also allowing viewers greater space for both cognitive and emotional exploration of a complex issue: what *is* to be done about the Ilva steelworks, which has offered so much employment in the region, and so much environmental and human health disaster? In suggesting an extreme possibility—destruction—the mapping thread, and thus by extension the film, offers viewers imaginative space to contemplate a future without Ilva, and perhaps the inspiration to work toward it via other, less extreme routes, as is called for by so many of the film's interview subjects. Above, I quote an email exchange with Zanotto in which she describes the directors' motivation for making this film. The fuller quote reads: "The idea at the heart of it was that that factory of suffering and pollution should be stopped, and our anger was strong enough to make it explode. And, in fact, in our own way we did!"[59]

COUNTERPOINT/CODA: BEYOND DOCUMENTARY

When examined together, *In viaggio con Cecilia, Buongiorno Taranto,* and *Non perdono* present a broad range of narrative approaches to the same ongoing crisis. This is most evident, as explored in the discussion above, in their unique framing conceits: *In viaggio con Cecilia* establishes viewers' sense of connection to one of the film's makers, so that we may join her in attempting to view the scene from the perspective of the other; *Buongiorno Taranto* focuses on the storytelling of a small collective, which seeks to elicit even more stories from the larger community; and *Non perdono* is based in a nonrealist frame tale motivated by affective response. All quietly innovative in their own ways, these three films are squarely place-based and address an ongoing real-life environmental crisis. In this, they remain largely within the traditional bounds of ecocinema, outlined at the start of the present chapter. They are not, however, the only films to have been shot in Taranto in recent years that might be examined from an ecocritical gaze.

Indeed, defining ecocinema as object of study, or category into which a film does or does not fall, can impose unnecessary limits and ignore a vast array of cinematic texts that might help us to analyze the more-than-human world and our relationship to it. A film like Michael Bay's action-packed *Six Underground* (2019), coproduced by Netflix on a $150 million budget, shot in a staggering number of locations worldwide (including the United Arab Emirates, China, Hungary, and the United States), featuring no discussion of environmental issues and certainly no long takes, also begs for ecocritical analysis. As Sean Cubitt writes: "The challenge for ecocriticism is not simply to identify and resolve a genre of ecological film. Or to analyze explicit ecological themes as they arise in film culture," but to instead "use its power to explain the absence of environmental issues, much as feminist critique did the structuring absence of women in certain films."[60] What's more, a film such as Bay's might prompt us to examine not only its storyworld, which features little apparent care for the nonhuman environment and almost no sense of place, but also to examine its means and process of production. *Six Underground* is, in fact and in part, also a "Taranto film," in that select scenes were filmed in and around the city's Città Vecchia. And yet only a viewer with inside knowledge of Taranto and its layout would know that, as no mention of the city's name is made in the film, and the scenes in question are instead presented as occurring in a vaguely defined "Italy."

If you do watch the film with an eye for the scenes filmed in Taranto in the summer and fall of 2018, or if you watch some of the ample set footage to be found online, you will note spectacular car crashes along the waterfront, flaming explosions at the iconic Sailor's Monument overlooking the Mar Grande and alongside a cruise ship docked just under the Ponte Girevole, as well as copious helicopters, cranes, smoke canisters, and crew members. This sort of large-scale action-packed filmmaking raises an important ecocritical question: what is the environmental impact of making film, especially on location? In her study of cinema's "footprint," Nadia Bozak reminds readers that "embedded in every moving image is a complex set of environmental relations."[61] Not only does the making and processing of film stock rely on a range of natural resources and chemical emissions (a material concern that is shifting to questions of energy usage and hardware production in this increasingly digital age), but the vast array of physical materials involved in supporting a film shoot use up additional resources and create permanent waste. Citing Corbett and Turco's 2006 report on sustainability in filmmaking, Elena Past writes that "the art, construction, set design, and paint departments leave 'tangible waste streams,' and large amounts of paper are often used for scripts and other photocopies," to say nothing of the daily residues, particularly plastic, left behind by catering. Such waste streams are often the result of an economically driven need for speed in production, what Past calls "a logic of immediacy," which can often override whatever interest in sustainability a film crew might otherwise possess.[62] To cite Laura Di Bianco, "a culture of speed and waste is profoundly rooted in the film industry because every minute costs money."[63]

In addition to what is left behind, we must also wonder about what is forever altered, about the structural impacts to local land- and cityscape of filming in a location such as Taranto. This is particularly the case for its evocative Città Vecchia, which is marked by narrow streets, ancient cobblestones, and close proximity to flowing waters. While the documentaries discussed above were not made devoid of people and equipment, they were made without constructed sets, large crews, or even scripts, thus leaving a significantly smaller footprint on Taranto. This is of course in keeping with the environmental awareness expressed within each film, as well as their attention to Taranto as an actual lived place. Past writes that filming on location "opens itself to a relationship with place and with matter, unleashing a creative force that offers ethical possibilities for envisioning an alternative, collaborative future."[64] With this she offers a challenge, to crews and to communities, to imagine a place-based filmmaking practice that gives space for local environments to take center-screen, that respects those environments' material conditions, and that recognizes the potential benefits to both place and people of hosting a film shoot. Rising to this challenge takes time and deep commitment, of the kind represented by the EcoMuvi international protocol for sustainability on film and audiovisual sets. As Di Bianco notes, EcoMuvi, which was launched by Italian-based Tempesta Films in 2014, "has made a tangible difference" in CO_2 emissions, use of plastics, and more, on the film sets for which it has been adopted. She also notes, however, that "creating ecological awareness in the film production process is an enormous collaborative endeavor," one that few productions have thus far adopted.[65]

While it did not, as far as can be discerned, adopt a particularly environmentally sensitive approach, a large production such as Bay's is no doubt positive for Taranto in an economic light, due to revenue gained by the city and local purveyors. I do not mean to dismiss this significant fact, as Bay's film, in collaboration with the Apulia Film Commission, which has helped finance a number of commercial film and television shoots in the region, offers just the sort of economic support that Taranto so desperately needs: completely unattached from Ilva.[66] And yet I do worry about the environmental impact of the equipment and materials required to film cars racing along the narrow lanes of the Città Vecchia and exploding along the Corso Due Mari, as well as the implications of filming in place without acknowledging place—using Taranto without telling any of its tales.

A similar sort of cinematic use can be found in *Belli di Papà* (2015, Dad's beauties), by the director Guido Chiesa. This mainstream Italian film is a comedy of morals, in which a wealthy patriarch teaches his children a lesson in humility by sending the family back to his native Taranto under the auspices of (false) bankruptcy. Significant portions of the film were shot in Taranto, which is in fact identified as such. And yet Taranto clearly stands as synecdoche for southern Italy, in largely negative and stereotypical terms that are only partially challenged: poor, dirty, backward, but full of kind-hearted "salt of the earth" people. The film offers no mention of Ilva or local crises of health, environment, and employment, and it provides only disjointed visual snapshots of the area. In notable contrast to the documentaries explored above, it actively prohibits viewers from engaging in any sort of imaginative mapping or developing a sense of place-based care, by centering action in dark interiors and, in rare external shots, filming protagonists at extremely shallow focal range so as to obfuscate background imagery. In terms of place representation, *Belli di papà* improves upon the model of *Six Underground* in that it does acknowledge Taranto as a particular named place, and yet it, too, uses Taranto more than it tells of it.

In closing, I mention a counterexample to the two titles briefly addressed above, as well as a possible model for engaging with Taranto as place—and particularly as place dominated by Ilva's ills: Sergio Rubini's *Il Grande Spirito* (The Great Spirit, 2019). Rubini's film is not without problems, most notably its embrace of racist tropes regarding First Nations cultures, but it strives to offer a model for recognizing Taranto as an actual living (and suffering) place that can support a fictional plot on-screen. It also offers a unique approach to mapping that puts its protagonist at eye level with the *mostro* Ilva. In brief, the film recounts the story of a petty criminal, played by Rubini, who, in his attempt to escape the former accomplices he has just swindled in a heist, winds up on the roof of a downtrodden apartment building in central Taranto with a seriously injured leg. He is found by a man (Rocco Papaleo) who lives in a crumbling shack of a space on the building's roof, identifies himself as Black Deer, and claims to be a member of the "Sioux Tribe of Canada." Black Deer cares for the reluctant patient, and, as the film progresses and viewers learn more of each character's backstory, an unlikely bond forms between the two. Notably, Black Deer has a history of mental illness and was raised by a steelworker father who used to come home covered in red steel dust.

As indicated above, the unblinking portrayal of Black Deer's (mis)identification as a member of an indigenous tribe and his adopted behaviors associated with that identity (song and dance, ceremonial fires, etc.) are quite problematic. What I wish to highlight for now, however, are the ways in which *Il Grande Spirito* acknowledges Taranto's complicated contemporary realities through both formal and allegorical means. On a formal level, the vast majority of the film's scenes take place on Taranto's rooftops: while the two protagonists are mostly shot atop Black Deer's building, viewers also see them navigate through the neighborhood by moving from the top of one building to another. This allows for sweeping panoramas of the city's physical layout, as well as a series of surveillance shots inside residents' apartments (summary: they all appear to be struggling). It also allows for repeated eye-level shots of Ilva's blast furnaces, directly parallel to Black Deer's rooftop. In fact, when the two men first meet, Black Deer gazes out at the smoking steelworks and decries the terrible destruction that has been wrought upon "his land" at the hands of "the Yankees . . . the white men . . . those who built the smokestack." He goes on to explain: "Here, before, it was full of fields, full of woods . . . it was so full of bison! Have you ever seen bison around here??" While the bison reference must correspond to his imagined past in Canada, it is true that the land where the steelworks now sits was once home instead to fields and woods. More significantly, the colonial metaphor in this establishing sequence, and thus throughout the rest of the film, is clear: the massive entity that is the steelworks has altered traditional lifeways and relationship to the more-than-human environment in Taranto, and forced residents into desperate living situations. When Black Deer dies at film's end, a sacrifice so that Rubini's character might live, Ilva's smokestacks notably complete the mise-en-scène.

Il Grande Spirito is a complicated and indeed problematic film meriting further discussion than what I provide here. It is also one example of how Taranto might function as cinematic protagonist in a variety of genres and methods of inquiry, even in narratives that are not wholly focused on the city's history or the impacts of the Ilva steelworks. I highlight this example, in contrast to *Six Underground* and *Belli di Papà*, to support a closing claim: when filmmakers choose to shoot in troubled terrain, they ought to acknowledge that trouble, and to lift up that terrain, by letting the film's place enter into its narrative. While the documentaries addressed earlier in this chapter offer strong models for doing so, just as they provide a platform for the stories of so many Tarantine residents, the films mentioned in these closing pages suggest that there remains more work to be done beyond the realm of documentary.

FIRST PERSON

Vincenzo Fornaro

I had the opportunity in the summer of 2018 to attend a day of the long-running "Ambiente Svenduto" (Sold-Out Environment) trial against former Ilva executives, accompanied by Daniela Spera. Sitting next to us in the court's otherwise empty audience chamber were Angelo and Vincenzo Fornaro. During one of many pauses in the day's proceedings, I slid down the bench for a brief interview with Vincenzo, who serves on Taranto's city council and ran a mayoral campaign in 2017 (he took 9.76 percent of the vote), in addition to managing operations of the family's land. As I explore further in chapter 6, the Fornaros now grow bioremediative hemp plants on the contaminated land where they once raised sheep.

MONICA: Can you tell me a bit about what you're doing with hemp?
VINCENZO: In terms of the hemp project, we're moving forward. This year we haven't planted yet. We'll try to plant at the end of September, and so we'll be changing the planting period, which is usually in March, to see what kind of results we can get. Anyhow, the climate in Taranto is particular. It could allow us to have a winter harvest. The month when hemp "shouldn't" be ready—we want to try to launch this. The idea is to start working with inflorescences [clusters of buds on a branch or stem]. Since there is no dioxin in inflorescences, it might be that we'll start there, a new project starting right from inflorescences.
MONICA: Is it legal for you to grow these crops on your property [due to documented soil contamination]?
VINCENZO: There is not actually a prohibition [on growing hemp]. It's just a question of understanding whether or not there is dioxin inside

the plant fiber, particularly the root. It could be that part of the fiber is contaminated, and so for precaution we've never explored [cultivating hemp for commercial use], to avoid any possible problems. But in the inflorescences there's no dioxin . . . so from there we could move forward, develop a business, and in the meantime see how the situation evolves. Unfortunately, the terrain continues to be contaminated, so every harvest becomes a risk.

MONICA: Do you check every time you harvest?

VINCENZO: Yes, every time. And we don't cultivate anything else, apart from olives, which we do harvest because it's been discovered that dioxin doesn't take hold inside the fruit; it just stays on the outer skin. You just have to wash every olive before it goes to press; then the dioxin doesn't pass into the oil. Olives are one of the only edible plants we're able to grow.

MONICA: And are you able to survive like this?

VINCENZO: Yes, yes. Not like before, of course. There's clearly not the same abundance as before. But we stay busy.

MONICA: What do you see here in five years? What do you hope for?

VINCENZO: My hope is that there will finally be a Taranto without Ilva: Ilva definitively closed.

MONICA: Is that possible?

VINCENZO: Of course! With remediation projects. All of the jobs that Ilva provides today have claims in exchange for a counterpart that Taranto can no longer afford to pay. Because the illnesses are still rising, and all of the nearby businesses around Ilva are closing up shop. It's a city that's becoming empty, that is aging because the young people leave, they don't see a future here in terms of job opportunity, or even the possibility of raising a family in this city, with such tragic levels of illness. It's obvious that young couples feel quite discouraged from raising a family here. And so the young people leave; as soon as they can, they leave: they go off to university, some don't come back, and some leave just to find a different job, outside of Taranto. So it's a city that is aging, demographically. When a city begins to age demographically it is, unfortunately, destined to die. Without a shadow of a doubt.

MONICA: So then how can Taranto survive without Ilva?

VINCENZO: Taranto can survive without Ilva precisely by hiring the steelworkers, plus additional laborers, to work toward environmental remediation. To dismantle this factory will take a minimum of thirty years. With funds from the EU and also establishing a policy of "those who pollute pay." There has been massive pollution here and now the time has come for those who have caused such pollution to pay for all the damages and take responsibility for the remediation efforts. . . . Taranto is so beautiful that it could profit from its great history and beauty and survive comfortably. When the city fills up [for events] everyone has work: the bars, the restaurants, the hotels, the taxi drivers . . .

MONICA: And what about the physical space of the steelworks . . . what do you imagine there? As an example, I was in Seveso last week to spend some time in the Bosco delle Querce, now a public park . . .

VINCENZO: That could be an idea! Leave just a few Ilva buildings . . .

MONICA: So that the history is still visible?

VINCENZO: Yes, yes. People could visit what Ilva once was, in the middle of Taranto. So, choose those structures that wouldn't leave a negative impact on the territory in terms of visual aesthetics or environmental harm. You could organize a sort of tour to see the change that has taken place, no? So that would be a kind of tourism too, an industrial museum. Many people don't know much about the steelworks. They haven't seen it. I went in with Daniele . . . it took us a year to photograph the whole thing.

MONICA: What is your goal in entering politics?

VINCENZO: [I'd like to draw attention to] the realities of common citizens, because so often we've seen that our realities are universally ignored by those in positions of power. So many of us, as active citizens, have decided to get politically involved because it's not right for our needs to be systematically ignored. It's very useful to get out in the streets, but after the protest there must also be a plan. We've proposed many plans, but they're always tossed out, so my goal was to enter the race and raise awareness, listen to the voices of the citizens, the actual active citizens with many different points of view. Active because finally it might lead to change, it might really . . . because otherwise we would still just be yelling in the piazza, which unfortunately doesn't effect change. But we want change, radical change. This whole situation . . . at a certain point we understood that we had to also get involved with local politics. It was no longer enough to be in the streets yelling out our arguments when we could instead be in a position to be heard by those with decision-making power.

MONICA: And for the book that I'm writing, for an English-speaking audience, what would you like me to write? What is the fundamental issue for you when someone writes about Taranto?

VINCENZO: To make people understand that at a certain point when . . . when we decided to rebel against the state of things, decided to denounce, to fight, we were rebelling against those who, instead, wanted to keep killing others, for their own interests. The interests of a few instead of the health of so many.

June 12, 2018

CHAPTER 6

Reading Landscapes
Back to the Land in Seveso and Taranto

Every time that I thought about the writing of this chapter, my brain conjured up a scene from a film that might not seem obviously related to the project at hand: writer and director Gianni Celati's *Case sparse: Visioni di case che crollano* (2002, Scattered houses: Visions of collapsing houses). In one subthread from the film, a meandering documentary that follows Celati and friends through the Po River valley in north-central Italy, a director coaches a lone actress as she practices lines for an upcoming performance about waning agrarian traditions. They rehearse in small-town piazzas, on benches in a bare train station waiting room, and along the banks of the Po itself. In this particular scene, the one that would come to mind when I thought of writing about land as text in Seveso and Taranto, actress (Bianca Maria D'Amato) and director (Alberto Sironi) stand outside on a wide set of cement steps connecting one street to another. They are continually interrupted: by cars speeding by, kids riding bikes or kicking soccer balls, the occasional horn blast, and so on. Sironi grows increasingly frustrated, running his hands through his hair. And yet in spite of the director's frustration, a relaxed feeling tempers the scene: the sun is shining, the slight sway of nearby trees conveys a breeze in the air, and D'Amato moves her body with playful ease, mildly entertained by the unanticipated exchange between people and place, the way real life continues to interrupt scripted performance.

The pair's chosen setting does not cooperate in the same way that the stage of a theater might, serving as static backdrop for D'Amato's physical actions and spoken narrative about what once was. Instead, it is place in its actual current state, interjecting its own actions and stories in a reminder that place remains, alongside the creatures that inhabit it, a dynamic actant.

In the language of Actor-network theory, and specifically in the words of Bruno Latour, an actant is simply "something that acts or to which activity is granted by others. It implies no special motivation of human individual actors, nor of humans in general. An actant can literally be anything provided it is granted to be the source of an action."[1] Some might argue that it is not place itself that serves as an actant in the scene described above but rather all of the many elements that the place contains: bikes, soccer balls, wind, etc. I counter, however, following Yi-Fu Tuan, that any given place contains and conveys spirit and meaning and, as such, can be conceived of as a unified entity, an amalgamation in which bike, soccer ball, and wind all play a part.

I suppose the scene from *Case sparse* came to mind for me when contemplating this chapter, which addresses current-day Seveso and Taranto as physical places with landscapes to be read, because in both locations place—and particularly the more-than-human nature that it contains—interjects itself in the narratives that people are seeking to recite. Place, in Seveso and Taranto, is both "acting" and "being granted activity by others," in ways that suggest complex new directions forward, alongside deviations from the script. As in the scene sketched out above, it is still unclear what the results of such acting / action might be—especially for Taranto, which is still living in the shadow of the *mostro* Ilva, as it tries on possible new identities grounded in tourism, urban expansion, artistic renaissance, and more.

In previous chapters I consider narrative texts that hold relatively traditional forms, primarily literature and film. I turn to studies in econarratology to consider the ways in which entry into the storyworlds contained in these texts allows receptive audiences to gain deeper knowledge about the effects of substances like dioxin moving through a local landscape, food chain, and human-animal community—ways in which, to cite David Herman, "states and events can be arranged into understandable and manipulable patterns; spatiotemporal relations can be established between regions of experience and between objects contained in those regions; relatively distant or intimate perspectives can be adopted," and more.[2] In what follows, I extend that approach in order to think through a treatment of land and landscape as texts containing similarly rich storyworlds. In the first section, I "return to Seveso," to cite again Laura Centemeri's study, in order to consider the Oak Forest (Bosco delle Querce), an expansive city park that has quietly flourished in the site of the former ICMESA factory for twenty-some years. Engaging the work of philosopher Eugenio Turri, as well as others who have written on walking as affective practice, from the aforementioned Tuan and Certeau to scholars of performance, body, and science and technology studies, I consider the patterns, relations, and perspectives to be found in the park. I seek to examine the story that it tells and to consider it as one possible solution to the question of what to do with a former site of industry and toxic exposure. With the Oak Forest still in mind, I then turn my attention back to present-day Taranto, reading sections of that landscape in a similar fashion and investigating in particular the ways in which a small group of residents and artists has begun to use plant matter for productive means.

This chapter is not intended to tie a neat bow on my study of Italy's dioxins and dioxin-tinged legacies, suggesting that all can be redeemed by reinvestment in place and land itself, or that nonhuman nature has the power to purify what has been greatly damaged if only we give it the space to do so. Rather, I seek to expand the consideration of narrative launched in previous chapters, reading landscape seriously as text or "storied matter," to return to Serenella Iovino's term. At the same time, I strive to offer readers a broader understanding of contemporary Seveso and Taranto, acknowledging that the projects underway there might just inspire a measure of hope.

In claiming to read landscape as text, I should clarify how I am using the term, especially in the context of two fairly urban environments, as well as what James and Morel describe as the "tendency in ecocriticism to perceive landscape as 'bad' or imperialist and environment as essentially 'good' or indigenous" and thus to devalue landscape.[3] A great deal has, of course, been written on the history of the term, which derives from the Dutch *landschap*, a close relative of the Danish *landskab* or German *Landschaft*. As Anne Whiston Spirn notes, the Old English version of the term, *landscipe*, makes clear the two roots involved: "land" ("both a place and the people living there") and the verb "to shape," what Marc Antrop instead identifies as "to make."[4] When we speak of landscape, we speak of the nonhuman environment in its overt shaping or making by human culture. This can be a physical shaping, as in modification of land by or for human use, including through the addition of built structures. It can also be a conceptual one, as through association with meaning, value, memory, and emotion—what Simon Schama refers to as "our shaping perception."[5]

Biggell and Chang point out that common understanding of the term "landscape" has widened in scope over the years, from "being a view or setting [that is, a distant something to gaze at] to being an arena where humans *interact* with the natural world." They also remind readers that landscape is necessarily limited, whether by the visual horizon "or by boundaries when we speak of social arenas."[6] Tullio Pagano similarly underlines the boundedness of landscape while citing the work of J. B. Jackson, who emphasizes its "synthetic" quality.[7] Pagano holds, furthermore, that "the 'constructed' nature of landscape is nowhere more palpable than in Italy, a country where virtually every inch of the land has been transformed to suit the needs of the inhabitants."[8] Indeed, Italy presents a particularly compelling case for the study of landscape, due to its long history of inhabitation, cultivation, and use.[9] Regarding the aforementioned tendency to consider landscape as inferior to "nature," it ought to be clear to readers by now that the present study recognizes nature and culture as deeply enmeshed and never truly separate from one another. Landscape is simply another iteration of this enmeshment, and I study it here for the complicated nature-culture story that it tells. In thinking about Seveso and Taranto, I am concerned with the general landscape of both cities—the land in and around those locations, as modified, thought about, and engaged with by its human residents. I am also concerned with individual landscapes within each location: the aforementioned Oak Forest, in Seveso, but also the land of the Masseria Carmine, the Fornaro family farm mentioned in chapter 3 that borders Ilva grounds, as well as the stretch of Taranto's coastline along the Mar Piccolo.

As noted above, I am often focused in the following pages on human interaction with the more-than-human landscape as carried out through the acts of walking or otherwise traversing the landscape in close physical proximity. The sensory awareness and kinetic coordination involved in walking, or in utilizing an assistive device such as a wheelchair, make for a necessarily engaged receptive experience, what Kimberly Powell calls movement as both ontology and epistemology.[10] In this, close movement through the landscape serves as a fully embodied form of reading. Spirn writes: "Landscapes were the first human texts, read before the invention of other signs and symbols. Clouds, wind, and sun were clues to weather, ripples and eddies and rocks signs of life under water, caves and ledges promise of shelter, leaves guides to food; birdcalls warnings of predators."[11] She lists "signs and symbols" that we perceive through the acts of seeing and hearing, but it is walking or moving through open space with mobility aids that carries the body through the landscape, bringing us close enough to the water to see the ripples and eddies, or to the trees to hear the birdcalls. Elena Past writes of walking as well, considering it a key part of her methodological approach in her place-based study of Italian ecocinema. She cites the work of ecophenomenologist David Abram, who "advocates strongly that we need to retune our senses and reawaken our awareness of our participation in the nunhuman world," just as he "contends that we must materially reconnect our flesh to the dirt beneath us, so that the bodily sensation restores our cognition of our co-constituted lives."[12] As Abram, Past, Powell, and others remind us, the experience of walking, to which I add the experience of moving in the open air with assistive technologies, draws our attention to the soil beneath us. And as I explore in this final chapter, it is soil that serves as the physical foundation for all landscape, and one of the primary realms through which the dioxins released in both Seveso and Taranto either bioaccumulate or, depending on neighboring substances, break down.

SEVESO: THE OAK FOREST

As outlined in chapter 1, the July 10, 1976, explosion at the ICMESA chemical factory on the Meda-Seveso border led to the immediate release of an unprecedented amount of dioxin in the form of a large toxic cloud. This cloud lingered over Seveso before eventually dispersing into the atmosphere, allowing its contents to settle onto solid surfaces, both vegetal and non, and to work their way into soil and water. While the exact amount of dioxin released that day has never been clearly established, an article published in the national daily newspaper *l'Unità* six years after the disaster gives some sense of the quantity, while also detailing a continued lack of concrete information in the disaster's wake.[13] The article's author, none other than Laura Conti, writes of a document released by a local magistrate in 1978 confirming that 2,900 kilograms of organic material such as carbon and hydrogen were released in the explosion, while another 2,300 kilograms of "carbonic residue," including significant amounts of dioxin, remained in the reactor.

Conti wonders why the court-ordered investigation that led to that 1978 document never determined precisely how much dioxin the material contained, as investigators had all of the necessary information to make such a calculation. She also notes, however, that they did not have access to the material itself. On September 10, 1976, just two months after the explosion, forty-one containers were shipped across the national border to an undisclosed location, by a private firm under the purview of Givaudan, the Swiss manufacturer that owned ICMESA. All told, they contained 2,200 kilograms of the contaminated material. In 1981, a media-led investigation found that the containers, which eventually numbered forty-two (a final container had been sent later), were being held in a small town in northern France. Upon release of this news, Givaudan's parent group, La Roche, retrieved the containers for incineration somewhere in Switzerland, still withholding the precise location. As Conti writes, "Givaudan produces the dioxin, Givaudan takes it away, Givaudan measures its concentration in the reactor, Givaudan decides where its remnants are taken and who is responsible for the transportation, Givaudan decides what information to give, and to whom; Givaudan does everything."[14] With this, she reminds readers once more of the lack of information and agency available to area residents (as well as to residents elsewhere along the containers' journey) regarding the harmful substances released, and perhaps still lingering, in their environs following the 1976 disaster.

Conti emphasizes a lack of information and agency regarding the 2,200 kilograms of matter shipped off to who knows where, but we must also wonder about the 3,000+ kilograms of matter that remained. While signaling an enduring frustration particular to Seveso, her article simultaneously speaks to broader questions regarding the correct steps to take when mass quantities of physical matter have been contaminated. Just what should be done with all the toxic stuff that results from a large-scale industrial accident—not only the most toxic chemical by-products released from reactors but the reactors themselves, the shuttered factories and nearby buildings, and the very soil resting underneath and encircling it all? Furthermore, how should any idling uncertainty or enduring resentment be addressed, especially when great unknowns remain? When it came to Seveso, that first question, involving physical matter, was resolved much more quickly than the second, which concerns the decidedly more complicated realms of information and affect.

Although communication was limited and inconsistent in the immediate wake of the disaster, both state and regional governments were relatively swift to begin cleanup and monitoring efforts in the contaminated zones. The most heavily contaminated, Zone A, was divided into seven subzones based on levels of dioxin detected in the soil, and on August 4, 1976, a special technical-scientific commission was created to study the case and propose solutions for decontamination. Under the guidance of Professor Aldo Cimmino, the commission eventually released two major studies, one covering the years 1976 to 1979 and the other covering the years 1980 to 1985. In the meantime, a number of important steps were taken regarding the future health of the damaged landscape: between 1976 and 1978 two state grants were issued to fund ongoing decontamination and remediation efforts, as well as chemical monitoring of soil, water, and vegetation, and in 1977 a series of regional bills were drafted that outlined the projected steps for long-term reclamation of the damaged land. The first phase of the plan addressed Zone B and subzones A6 and A7, for which it proposed defoliation, disinfection, removal of topsoil, isolation of heavily contaminated tracts of land, identification of materials destined for incineration, and eventual remediation of remaining plants and soil. The second phase of the plan, dedicated to the more heavily contaminated subzones A1 to A5, outlined demolition of buildings, removal and surveying of topsoil, investigation of possible solutions for disposal, and the eventual construction of a large incinerator to burn up all of the heavily contaminated vegetation, building material, and carcasses of all those animals that had been slaughtered due to contamination concerns. The goal was to achieve complete cleanup and incineration by 1980.

While most of the aforementioned steps were carried out, the contaminated matter was never fully removed from the site, and the incinerator never came to be. As initial decontamination efforts progressed in subzones A1–A5, and as cleanup crews prepared to dismantle the toxic reactor, they were unable to identify any Italian regions or neighboring countries with existing high-temperature incinerators or ideal dumping sites willing to take on Seveso and Meda's toxic stuff (a chronicle of events from the Lombardy regional offices lists "abandoned salt mines" as a possible example of such sites). Furthermore, despite the insistence of Professor Cimmino and his commission regarding the need to completely destroy all of the matter in question, local authorities and the community at large voiced strong opposition to the building of an incinerator. They feared the land that once housed a chemical manufacturing factory would now become a permanent disposal site for an even vaster array of toxic matter, as the incinerator was slated to serve all of Greater Milan. Not only would it be a financial burden to maintain such an operation, the idea of housing an incinerator felt to the residents of Seveso and neighboring communities like insult added to injury, as they would inevitably be exposed to further harmful emissions.

Centemeri notes that for some residents the battle against the incinerator was primarily driven by a technical question (how efficient and safe would it truly be?), whereas for others it was tied to something deeper. As she writes, those residents felt invested in the "affirmation of the landscape as a patrimony to conserve and transmit to future generations. A landscape in which the characteristic trait was found in the presence of ample wooded areas."[15] Motivated by sense of place and respect for a plant-based patrimony, they fought not only to prohibit the incinerator but also to restore Zone A to the sort of verdancy it must have boasted before ICMESA was constructed in the 1940s. In response to such efforts, the region approved a new plan in 1979 to transform part of Zone A into a protected green space. The surprise twist: two toxic landfills would be buried deep within its very center.

As remediation progressed, all contaminated materials were temporarily stored in secure containers and new soil brought in to begin building up the excoriated land. Then, starting in 1981, two massive underground vats were constructed at the site of greatest exposure, right where the towns of Meda and Seveso meet. Each vat was built of four successive material barriers: the first two were made of organic matter, sandy and clay-heavy soils known to absorb dioxins, while the third and fourth were made of polyethylene and a mixture of compacted inert matter such as bentonite, an absorbent mineral-rich clay. Once the vats were full, they were covered first by an additional layer of polyethylene, then by one of concrete, and connected to a complicated drainage and filtering system.

Completed in 1984, these vats present a fascinating object of ecomaterialist study. As Massimiliano Fratter writes, "most of the contaminated material is made up of the top level of soil from the most contaminated area (Zone A), along with parts of houses, machinery, personal objects, dead animals, and some of the initial cleanup materials."[16] In their mix of contents, the vats offer a clear representation of our deep enmeshment with the surrounding material world. Containing the human-made objects of intimate daily existence alongside bodies of animals, mechanized devices used to produce both perfumes and herbicides, and so much dirt, they offer a sober tableau of what Jeffrey Cohen describes as "life in a vortex of shared precariousness and unchosen proximities."[17] And in their multilayered barrier system, comprising minerals, clay, plastic and concrete, they emphasize the potential for this vortex to cause harm, to necessitate containment and isolation when things get out of whack.

Particularly intriguing is the presence of soil in both the contaminated mixture within the vats and the material barriers encasing them. The topsoil resting beneath the ICMESA factory and surrounding land was exposed to high enough levels of dioxin so as to support bioaccumulation, an increasing of the compound through the plants and animal bodies it would, in time, enter. This soil was toxic. And yet soil can also help to eventually degrade toxic compounds, through processes of oxidation and metabolic breakdown performed by other microorganisms. The Soil Health project run by Lyn Abbott's team at the University of Western Australia notes: "The ease with which a chemical is degraded depends on the complexity of its structure. More complex structures are more difficult to degrade because fewer microorganisms in the soil produce enzymes capable of degrading them."[18] Although dioxins do present fairly complex structures, Seveso researchers found that approximately 95 percent of the dioxin released from the ICMESA explosion rested in the top twenty-five to thirty centimeters of surrounding soil, and that the clay-rich substrata beneath served as an effective absorbent for what remained.[19] As such, the top forty-plus centimeters of soil were removed, but lower levels were allowed to stay put.

The humblest of materials, soil is in fact quite potent. It has the ability to transport toxic substances, and it has the ability to dismantle them. As those underground vats suggest, in comprising the very ground beneath our feet and the matter in which our food grows, soil does the same for *us* as well, both transporting and dismantling (as Past reminds us, "dirt [soil] is important everywhere").[20] Thinking in terms of Mel Chen's work, and particularly Chen's scale of animacy as discussed in chapter 4, soil—like dioxin—is an actant containing both "subjectness" and "objectness," and an active part of the human-nonhuman mesh that we call landscape. Fittingly, soil is a key component in the next movement of the Seveso story.

Just as complex as those material-rich toxic vats is the park that was soon built directly on top of them. During the same time that the vats were being constructed, local officials moved to begin turning the land under which they rested into a public greenspace. This decision came largely in response to the aforementioned popular movements opposing the originally planned incinerator, which sparked an environmentalist wave. The space designated for the project consisted of approximately forty-three hectares of land that had been scraped entirely bare, with the exception of one lone remaining tree: a poplar, a symbol of strength and resilience due to its deep root system. Construction of the park officially began in 1984, when massive quantities of soil were brought in from other communities within the region—a practice not unlike that of blood transfer when a living creature has been gravely ill—and careful planting first undertaken.

In his detailed history of the Oak Forest, Fratter describes it as "a forested area of anthropic origin, with a vegetal composition inspired by the original woods nearby which are filled mostly by English oak, sylvan pine, birch, white hornbeam, black alder and white willow."[21] As he explains, the park grounds alternate between grassy field and more densely wooded areas, including a particular section of "woods of spontaneous evolution, characterized by a rich underwood, including a limited access area for naturalist vocation."[22] Walking along the park's paths, he writes, one can hear the song of woodpeckers, great tits, blackcaps, winter wrens, wrynecks, hedgehogs, tree frogs, shrikes, and European warblers. There are other creatures as well, such as butterflies and rabbits, along with flowering plants like dandelion and verbena. Fratter also explains that this verdant universe, thriving in the site of former chemical production, was only slowly opened up for public enjoyment, and not without some difficulty.

Although the Oak Forest was officially complete in 1986, it was treated as an experimental remediation project for the first ten years of its existence, during which time the public was welcomed in on very rare occasion. This was due in no small part to lingering concerns about dioxin concentrations in the immediate environment. The park was declared safe for public use in 1991 following investigations by the office of the regional environmental assessor, but opened to the public only in 1996, and then only on Sundays. Public use was eventually expanded, and the park is now open seven days a week, May through September. The Oak Forest has continued to evolve since its initial planting, with new species of plants and creatures spontaneously making their debut, and now hosts children's summer camps and a wide variety of family-centered programming during the weekends.

Since the early 2000s, the Oak Forest has also housed the collective memorial project Il Ponte della Memoria, the Memory Bridge. First conceived in 2001 by the Legambiente Lombardia, in collaboration with the City of Seveso and the Fondazione Lombardia per l'Ambiente, the project was intended to offer the local community a range of archival materials through which to conduct historical analysis and cultivate collective memory of the

Seveso disaster. As stated on the Oak Forest's official website, the Memory Bridge was explicitly inspired by historian E. H. Carr's reminder that "history is an unending dialogue between the present and the past."[23] It includes a rich archive, didactic programs, and what it calls the Percorso della Memoria, or Memory Path, a series of twelve large panels situated throughout the park to share images, information, and first-person testimony. The panels lead park-goers through areas for play, sport, and continued bioremediation, as well as open green space where they can lounge on the grassy knolls covering the twin vats.

At first a collaboration between the groups mentioned above, the initial archival project expanded its reach in 2003 to include contributions from the Fondazione Corriere della Sera, which contains the eponymous newspaper's archives, and was eventually entrusted to the local Legambiente chapter around the same time. A team of members from the latter worked together to compile documents first from newspaper archives and then from community residents themselves. These documents included articles and previously published photographs but also personal letters, handwritten signs of warning or protest, and oral interviews conducted by the Legambiente team. The team compiled the materials to be stored at the park in an on-site archive and to serve as source material for the twelve panels, which would feature both narrative text and images. As work on the panels progressed, an additional team of sociologists, psychologists, and select community members was assembled to reach out to the broader community for input and involvement. They facilitated a series of town hall–style meetings with residents in the hope of creating a piece of shared writing (*scrittura condivisa*) to constitute the panels' text.

The psychologist Stefano Carbone recounts the team's desire to consider the meetings a "symbolic funeral" for Seveso, providing residents space to share thoughts and feelings about the disaster and its aftermath: "We had thought of a project that would give people the opportunity to release their anger, their pain. Because there in Seveso one feels that there's something in the social body . . . something that's not okay." Despite such intentions, however, community members were not very willing to engage: "There was no budging: it was clear right away that there are certain things one shouldn't speak about."[24] As a result, the team opted instead to represent what they called a "discrete" memory on the panels. This memory focused on the community's resilience and dedication to environmental remediation in the decades following the disaster. It did not, as Centemeri points out, address "the thorniest question—the damaging health effects *in primis*," nor did it do much to address the conflicts surrounding communication, abortion, or eventual financial remuneration.[25]

Centemeri notes, further: "The collective memory that takes shape in the panels of the Oak Forest insists upon the capacity that the residents of Seveso showed to put the disaster behind them, but it ignores the questions that divided and still continue to divide. Based in a sort of positive vision of the incident, it tries to fit the disaster into the formation of a local identity that valorizes associationism, rootedness in the landscape, and care for the environment, all at once."[26] In sum, the twelve narrative panels that resulted from the community outreach tell a story of the disaster and its aftermath, emphasizing a spirit of community resilience rather than more complicated affective experiences such as doubt, resentment, shame, or anger. I briefly mention the struggles faced in creating the panels, and cite Centemeri's critique, to underscore that the crafting of historical narratives inevitably involves decisions about what is included and what is left out, and that this becomes an especially delicate task when seeking to represent a collective voice. Although an experience may be shared, the responses and interpretations that it provokes are rarely identical. It is for this reason, I argue, that many contemporary projects of counterhegemonic storytelling are focused on recounting a multitude of individual voices, such as in the Taranto texts addressed elsewhere in this book.

In 2015 I visited with Fratter in Seveso's Oak Forest. We spoke about the park's history and current programs, which feature narrative-based didactic activities to teach children about local flora and fauna (Fratter has also authored a series of short stories featuring a cheerful frog that are in heavy rotation). An original member of the team responsible for the Memory Bridge project, he spoke openly about the team's motivations for their work, as well as his own mixed feelings in the years since. Speaking of Seveso at the time of the disaster, Fratter echoed both Centemeri and Conti. He said that in contrast to the image presented by the panels, "there was not one vibrant community, but many small communities: the Catholic group, the early environmentalists, Laura Conti . . . but there was not cohesion between the different positions. In fact, there was polarity." And yet, in the early 2000s, the Memory Bridge team agreed that minimizing a sense of community discord within the panels' narrative "was a necessary mediation, even just to bring people back to the park."[27]

This final sentence is, I believe, fundamental. The creation of the Memory Path within the Oak Forest was motivated by two primary goals: to nurture collective memory of the Seveso disaster and to offer its story as a lesson for contemporary park users. As Centemeri's study and Fratter's reflections suggest, that first goal may not have been successfully implemented. In this, the project can serve as a negative model, a lesson for current and future communities struck by environmental disaster, such as Taranto, as they contemplate possible means of public remembrance and memorial. The second goal, however, has been largely met. When a visitor to the Oak Forest reads the Memory Path's narrative panels, they encounter a clear physical history of the land on which they stand, through the panels' descriptions of the explosion at ICMESA, the extreme quantity of dioxin released that day, and the tremendous amount of work that went in to decontaminating the surrounding landscape.

In 2004, when the then-mayor of Seveso, Clemente Galbiati, inaugurated the Memory Path, he noted that without the panels a visitor could enjoy the Oak Forest and leave with no knowledge of what had previously existed in that same site: "Nothing is out of place, everything is 'perfect.' Those who don't know the origins of the area are struck by the beauty of the zone."[28] By adding the narrative panels, and by placing them throughout the park, the curators of the Memory Path brought the land's history overtly back into the story that it tells on any given day. Gemma Beretta, longtime director of the local Legambiente chapter and another original member of the Memory Bridge team, reiterated this point when I spoke with her and colleagues shortly after having first met Fratter. She explained that when the Oak Forest was planted and first opened to the public, "there was no trace of memory of what had happened in those woods," and so "we thought something was taking place that wasn't okay, that . . . a situation was going to be normalized without any memory of where that environmental remediation had come from, of what is there *underneath* those woods."[29]

Beretta's words, alongside those of Galbiati, call us back to the fundamental role of landscape in the larger story of the Seveso disaster, and to the humble matter that is soil. The narrative that unfolds on the Memory Path's twelve panels may fall short of capturing the complex workings of a wounded community. By overtly signaling what the land once was, however—by causing park users to pause, to look around, to imagine that a factory once stood there and that one day it witnessed a chemical explosion—the signs give park users the primary push we need to recognize the wounds inflicted upon the land itself, as well as its multilayered history. They cause us to become active and engaged readers of the land beneath our feet, ultimately more content-rich than the signs themselves.

My first opportunity to read the landscape of the Oak Forest came that same summer in which I first met with Fratter, Beretta, and others. I had rented a house for ten days in Cesano-Maderno, three miles south of the former ICMESA site, in what had been labeled Zone B almost forty years prior. Each day of my visit I rode a commuter train for a brief five minutes before disembarking in central Seveso and walking up to the former ICMESA site. The town itself was generally quiet, hazy in the light smog that seems to hang around Milan and environs in the summer, and home to a compact downtown area surrounded by modest but well-kept single-family homes. On my first pilgrimage to the park, I was surprised to find the Oak Forest nearly unmarked, a single open gate offering a quiet invitation just past a cemetery, indoor bocce ball courts, and what looked like a few small manufacturing sites. As I was to learn later, I had stumbled across a back entry: the main entrance sits just off of a pleasant residential thoroughfare and is decidedly more pronounced, although still far from ostentatious. Upon first impression, the Oak Forest appeared to me like so many public parks in Italy: a bit of trash and graffiti around points of entry, followed by an open stretch of lush green grass dotted with ample shade trees. The Oak Forest is shaped like a long, thin triangle, bordered along one edge by the old state highway that leads from Meda to Milan, with a gravel pedestrian path winding through its middle. As I began to walk through the park, I noted various other users—cyclists, dog walkers, teenagers chatting and studying for exams—as well as benches and flat, wide informational signs, those Memory Path panels, scattered throughout.

Before leaving for Italy that summer, I had asked a geographer friend who runs the Parks Research Lab at my university what I ought to look for when considering the Oak Forest as a text. She told me to pay attention to three things: views, paths, and signage. Where, she said, is your gaze directed? What do you notice, and how do physical elements within the park guide that? A few days into what became my daily park routine in Seveso, it struck me that I only saw the park itself. That is, I was visually immersed in greenery once I had wandered down the path a bit, no matter which entrance I had used. The Oak Forest sits slightly lower than surrounding areas, so that it felt as though I were tucked into a cozy nest of green. When I looked around, all I saw were grassy field space and leafy trees rather than smokestacks, billboards, or neighboring homes. The gravel-lined path was fairly narrow, wide enough to allow multiple park users to comfortably pass by one another, but otherwise demurring to the surrounding verdancy. It took a slight curve as it led from one entrance to the next and peeled off toward both ends, inviting users to access a playground tucked back into one corner, as well as a rather overgrown pond behind a thick hedge that is intentionally left uncultivated.

Small signs note that this pond, one of a few designated "naturalist" areas tucked along the park's outer edges, has only recently been opened to the public. Posted signs instruct visitors to remain on the small outlined path and not wander farther into the brush, so as not to disturb the activity of the native micro- and macrofauna growing there, as well as the animal species that now call it home. This is not exactly an example of rewilding at work, as the areas in question are relatively small (a total of 16.5 hectares spread throughout the park) and certainly not home to large predators, but they do offer an example in miniature of "restoring natural processes and species, then stepping back so the land can express its own will."[30] I was acutely aware of being a visitor there, a minority subject in an area dedicated to the experience and agency of the nonhuman.

The rest of the Oak Forest is more cultivated, however, and a particular kind of landscape in that it is also a park, situated not far from industry, dense housing, and city center. Sadeghian and Vardanyan define a park as "an area of natural, semi-natural, or planted space set aside for human enjoyment and recreation or for the protection of wildlife or natural habitats," and note that parks are "geared towards fulfilling the leisure, recreational and educational needs of the young and old, male and female, rich and poor, and of people of varying abilities."[31] J. B. Jackson, in turn, writes of a park as a "public, open-air space where we can acquire self-awareness as members of society and awareness of our private relationship to the natural environment."[32] Whereas the first definition emphasizes a rather anthropocentric democracy, albeit recognizing wildlife and their habitats, the second presents an intriguing blend of self-reflection regarding one's relationships to both the human social sphere and the nonhuman environment. It is this second understanding of a park that I found promoted in the Oak Forest, through the copresence of spaces like the naturalist areas mentioned above and the twelve panels of the Memory Path.

Three copies of the Memory Path's introductory panel greet visitors throughout the park, one at each of its three entrances. They are placed in such a way as to not be missed, so close to the footpath at the primary entrance and a few feet opposite the open gates at the other points of entry, that you have to go out of your way *not* to read them. This is just where one might expect a general park overview sign to be, in the sorts of designated nature spaces that tend to feature signage. Think, for example, of your local hiking area or city park built in a site deemed historically significant. Where I live, a North American city that was once the capital of the Confederacy, the riverside trail system that runs through the middle of the city features a number of wrought-iron signs offering a brief political history of the space (once a site of battles, armories, prisoner-of-war camps, and so forth), information about native species, and codes of expected conduct for park visitors, such as not littering, minding dogs, and leaving at dark. The first panel of the Oak Forest's Memory Path does something similar. In a long column of printed text next to striking photos from the Volpi family archive, the panel welcomes visitors and situates the space geographically, before offering a concise history: "The Forest was born in 1983. Environmental and forestry work began in 1984 and ended in 1986. The Oak Forest was planted on land heavily polluted by the toxic cloud ('Zone A') that emanated on July 10, 1976, from the Meda ICMESA factory and that contained Dioxin (2,3,7,8-TCDD), among other toxic substances."

While the form does not exactly subvert our expectations, the panel's content alerts visitors to the fact that the Oak Forest holds a rather unique history for a park. In this alone it might snap us out of our normal habits, which tend to feature a distracted sort of reading. What is even more particular, however, is the final section of the panel, where we might expect a code of conduct to appear. After a few sentences explaining the decision to turn the land into a park and underlining its symbolic importance, readers find a final brief paragraph. Rather than hours of operation or rules of use, we read: "The Oak Forest is a living place with a new story to tell. Along its paths, an itinerary has been erected that helps to not lose the memory of what occurred, an environmental and human wrong, curated thanks to the many acts of men and women motivated by a profound love of the land from which they come." Although no rules are spelled out or commands given, this language invites park users to be a receptive audience by following the progression of the land's story, while remaining mindful of the significant wrong (in Italian *danno*) that informs it. Explicitly referencing the Oak Forest's story, and thus its narrative agency, this introductory panel encourages visitors to be careful readers as we move through the landscape.

In contrast to the introductory panel, the Memory Path's remaining eleven panels are set farther back from the gravel path that runs through the Oak Forest, nestled unobtrusively against trees and shrubs. To read their content, perhaps also examining the large photographs they display, one has to step off of the footpath and wander through the grass, feeling the land beneath their feet and moving nearer to the trees and shrubs by which the panels are erected. Flat, muted, and held up by a wrought-iron frame, they do not visually dominate the park, but they do mark it as both spatially bounded and temporally steeped in human use. What's more, they encourage users to actively engage with the landscape rather than remaining within scripted territory as static observers only allowed to move on a designated trail. Stepping onto the grass to approach a panel, we act as dynamic subjects ourselves, taking part in the current landscape as we choose to read about its past. Reading printed text such as, "This hill, located in the middle of the Oak Forest, is the so-called Seveso Tub, one of two tubs constructed for storing and securing the contaminated materials. . . . Not only soil. Also contained in the tubs are the memories of those who were forced to leave, starting on July 26, 1976," we are still receptive to the landscape. We might note the scent emanating from a nearby tree, feel the grass against a bare ankle in warmer weather, hear a critter moving through the underbrush, all as we imagine what came before in this same space.

In my first visit to the Oak Forest I found this to be a moving experience, but I recognized that I had come to the park with my head already full of the texts I had read about its history; I was primed to be a particularly receptive reader. In 2018, I returned to the Oak Forest, curious about how others interacted with the space. As during my previous visit, I was able to access the park for a few days in a row, strolling both on and off the footpath, rereading the signs along the Memory Path and trying to observe the habits of other park visitors. In my very casual study, I noticed people really making the most of the park, especially on weekends: adults pushing toddlers in strollers, lots of cyclists, groups of teenagers lounging in the grass, solo runners, older women huddled together on benches tucked under the trees, you name it. This was heartening, but I had to wonder, does anyone read the signs? When people come to spend a Saturday afternoon in the Oak Forest, most of them probably repeat visitors from the local community but some from beyond it, how aware are they of its historical layers? Of the ecological catastrophe that made this space of leisure and recreation possible? Of the massive vats of toxic waste underneath the imported soil that supports that soft green grass? And if they're not reading the Memory Path's narrative panels, how are they reading the landscape?

I never did approach those park visitors to ask them all of the above, but I did get one helpful answer. Stationing myself on a bench near the park's main entrance a few afternoons in a row, I confirmed my earlier hypothesis regarding the introductory panel: most of the people who entered did pause there, some skimming the panel rather quickly as though already familiar with its content, others lingering over each word. I am convinced that both its physical positioning and surprising story are crucial factors in getting park visitors to read the Memory Path's first panel.

As Hall, Ham, and Lackey report, we are much more likely to read park signage if a sign contains novel designs and the use of "vivid information," including visual imagery and first-person narrative. Referring to an earlier park-use study by Baesler and Burgoon, the trio writes, "The vividness of a story and the structure of its components engage attention without requiring concentrated effort on the part of message recipients."[33] If we return to the definitions of "park" quoted earlier, we recall that parks are generally spaces of leisure, not arenas for concentrated effort, unless that effort is directed toward physical recreation. Authors of narrative signage are thus confronted with the challenge of engaging potential readers, but not too much. This raises the question, of course, of what can be gained from reading in such a way, as well as the question of whether a "leisurely" reading is possible when the content is so serious. Are readers able to enter into storyworld experience without concentrating? Can they gain enough cognitive and affective awareness to care about the land on which they stand or rest, through what might only be a cursory read? I believe that the answer is yes and that this is primarily thanks, in the case of the Oak Forest, to the experience of embodied reading that accompanies the narrative signage—the experience, that is, of reading the landscape itself.

After the first strategically placed panel, visitors can choose to walk or otherwise move through the park without reading any further, but we cannot read any further without walking or otherwise moving through the park. That is to say, we can read all of the panels and thus allow the printed narrative to guide our embodied experience of the Oak Forest's particular landscape, or we can read the introductory panel and keep that first capsule story (1976 factory explosion, dioxin, remediation, a living place with a new story to tell) in the back of our heads as we move on to our leisure activities. Either way, we traverse the landscape with our bodies, gathering information sensorially through movement and physical sense like sight, smell, and hearing. We might note some of the plants and animals that flourish there, just as we might note that its verdant green eco-system is in stark contrast to the surrounding residential neighborhoods and eventual commercial centers. With even just a minimal knowledge of the ICMESA explosion that occurred in that same site decades prior, we might read the landscape before us as indication of the ways in which humans can help support the agency of nonhuman nature, as well as a reminder of the ways in which we can gravely harm it.

Such awareness echoes the philosophy of Eugenio Turri, who argues that landscape is a theater wherein the performance of nature-culture progression takes place.[34] Much like the scholars cited at this chapter's outset, Turri's approach to landscape is overtly anthropic: he sees it as a space that inevitably bears the sign of human intervention and, as such, provides visual cues to human history. He writes: "One way to take on the natural condition in reading the landscape, to assume it as a fundamental component . . . would be precisely to consider the human sign as the result of a communicative relationship between man and natural environment."[35] In addressing nonhuman nature by way of the human, Turri establishes a hierarchy of value in which the human clearly reigns. And yet by emphasizing the "communicative relationship" between human and nonhuman nature, he describes a mutuality not unlike that explored in the work of contemporary eco-materialist scholars.

Turri is interested in landscape for what it tells us about human history, but he is deeply attentive to the role of nonhuman nature in writing that history—its vibrancy, its agency, its animacy. As a result, he encourages us to pay very close attention to the landscape, to read it, and declares that even our leisure walks, what he calls "our walking through the hills," ought to take note of the wider spatial and temporal dimensions in which local landscapes insert themselves. In his attention to close reading and his reminder that, much like "man and natural environment," past and present are ever entwined, Turri's work recalls that of E. H. Carr, from which Seveso's Oak Forest takes its inspiration. Turri proposes that spending time in the landscape simply for the sake of being there, "breathing its breath," engaging in the practice of walking, allows people to be both spectators and actors at once, or better yet, "active, non-inert spectators." In a longer version of the passage referenced above, he writes, "The local space, the minimum territory of our walking, inserts itself in a greater spatial and temporal context—of which even our leisure walks should take note."[36]

By quite literally signposting the Oak Forest's multilayered history, the Memory Path project—at least its first narrative panel—heeds the call issued by that verb "should" (in Italian: *dovere*). It encourages visitors to be active readers of the landscape as we move farther into the park, whether or not we continue reading the remaining narrative panels. It also encourages visitors to recognize our own position within a multilayered history. Citing the work of Catriona Mortimer-Sandilands, Iovino writes: "If remembrance is 'a recognition of a relationship between the body/mind and the external world that is not only determined by internal forces' (Mortimer-Sandilands 2008: 274), then landscape is the deciding site where the relation of inside and outside, body/mind and world, gets reinforced or progressively erased."[37] By physically signposting the complicated history of the Oak Forest, and the toxins once emitted and still contained in that space, the Memory Path overtly prohibits its users from such acts as erasing or forgetting.

As in the scene from Celati's *Case sparse* described at the outset of this chapter, things don't always go according to script. At the time of the Memory Bridge's formation, the Seveso community may not have been ready to engage in a "symbolic funeral," releasing all of their frustrations surrounding the ICMESA explosion and its aftermath; and still today, visitors to the Oak Forest may not read all twelve narrative panels of the Memory Path. Yet if we enter the park as both actors and spectators, aware of the landscape's history as we simultaneously read and take part in its ongoing present, we can still learn so much. In this, Seveso's Oak Forest offers a positive model for current and future communities struck by environmental disaster, such as Taranto, as they contemplate possible means not only of public remembrance and memorial but also of remediation, regeneration and a more nuanced "communicative relationship between [hu]man and natural environment."

TARANTO

At a small symposium a few years back, gathered together around a dinner table following a daylong exchange of stories and ideas, some colleagues and I decided to propose a panel for a much larger conference the following year. The panel, we determined, would highlight our shared interest in Italian landscapes and texts, as well as our shared belief that they matter to the environmental humanities at large. To guide us, we settled on three key terms. "Landscape" would surely be one, and "resistance" was an easy second, borrowed directly from the conference's official theme.[38] The final term, however, took a bit more brainstorming. Ultimately it remained in transition, mutable as landscape itself: both "performance" and "performing" were to lead us, noun (text) and progressive verb (action). The resultant shorthand title, "Performing Resisting Landscape," made space for our individual explorations of contemporary Italian landscapes as epicenters, rich with story, of a centuries-long back-and-forth between nonhuman and human ways and uses. It also spoke to our shared desire to move beyond what had been for all of us a focus on more traditional narrative texts, such as literature and cinema.

For me, "Performing Resisting Landscape" offered a chance to think further about the physical terrain at the root of my ongoing study of Italian narratives of toxicity, which was just beginning to take shape as this book. It also provided the opportunity to explore a set of human interactions with that terrain that are uniquely performance-based, temporally expansive, and guided by courageous optimism in the face of great devastation. With that in mind, I intend for those same terms, selected in good company on an early-winter evening, to guide my thoughts for the rest of this chapter, in which I consider a set of activists and artists performing both resistance and recovery in Taranto by directly engaging the landscape itself. The aforementioned Fornaro family, the artist Noel Gazzano, and the Ammostro arts collective all work in various ways with Taranto's material landscape, taking stock of its past and looking toward potential futures. Whether through walking, planting, or creating material objects, they invite others to join them in their practice. In this, they draw participants into an embodied awareness of the continual transition both experienced and offered by the soil beneath their feet, surpassing even an active reading of Taranto's landscape, in order to participate in its performance.

While the Fornaros, Gazzano, and Ammostro all focus on Taranto's more-than-human environment in their work, the first two are especially linked through the act of planting hemp, the nonpsychoactive strain of the plant *Cannabis sativa L*. Hemp, their work suggests, might provide recovery and regeneration to Taranto's land and creatures in two distinct ways. The hemp plant is easily cultivated, and, as Andre, Hausman and Guerriero note, it is also a sustainable resource that makes good climatic and economic sense, in that it provides a "source of fibers, oil and molecules and as such it is an emblematic example of a multi-purpose crop."[39] In promoting the cultivation of hemp on their family land, the Fornaros emphasize that the plant's fiber can serve as raw material for both textiles and ceramics, and thus a regenerative source not only for Taranto's historically agrarian ways but also for the region's traditional forms of productive craftsmanship. They and Gazzano also stress the hemp plant's bioremediative properties, its ability to break down or consume environmental pollutants in a given site. In a poetic twist, hemp shares certain characteristics with dioxin, in that it is a transmutable, transcorporeal organic substance able to change the makeup of soil—in this case for the better.

This process is known as phytoremediation, the use of living plants to break down the contaminants in soil or water.[40] Phytoremediation can take place at multiple stages of a plant's growth cycle and at multiple levels of interaction with the substance in which it grows. A remediating plant might degrade contaminants within its tissue; transform or volatilize contaminants into the atmosphere; translocate contaminants from soil or water and store them in its roots; bind contaminants into soil and thus prevent them from moving on; or manipulate "rhizospheric associations between plants and symbiotic soil microbes to degrade contaminants."[41] Since the 1990s if not earlier, various mushroom, brassica (mustard), helianthus (sunflower), and other plants have been used for phytoremediation projects, as has hemp, most notably at the site of the 1986 Chernobyl nuclear disaster.[42] Carolyn Beans writes that in recent years scientists have "steadily advanced phytoremediation technologies, coaxing plants to detoxify a range of pollutants ranging from lead in abandoned mining areas, to pesticides on old orchards, to petroleum hydrocarbons resulting from gasoline leaks," but the practice is "still struggling to make the leap from lab to field."[43] Studies suggest that although cells throughout the hemp plant will hold toxins, it decontaminates soil primarily through its roots, which can tolerate a high level of toxicity while still producing robust leaves and stems.[44] In this it is a particularly well-suited plant for Taranto, both physically and symbolically, as it grows strongly toward the future.

I have already briefly mentioned Angelo Fornaro, with whom I sat for a day of the Ilva trials in the summer of 2018. Born in 1934, Don Angelo, as he is known in the community, is the lead proprietor of the Masseria Carmine, the Fornaro family's large farm just abutting Ilva property lines.

A kind and soft-spoken man, Don Angelo has a deep love for the more-than-human world, from the many horses his family keeps to the many hectares of land on which they roam. In an article written in honor of his eighty-fifth birthday, his daughter Rosanna shares: "Dad has always loved, and still loves, the land, the olive groves, the animals. . . . [H]e especially loves white farmhouses, walking around the grounds, the thought of raising animals, and of managing the operations well."[45] It is for this love that Don Angelo has attended nearly every day of the Ambiente Svenduto trial against former Ilva executives. On the day that I attended, accompanied by Daniela Spera, Don Angelo leaned over to speak with me for a bit during one of the countless pauses in proceedings. In a near-whisper he explained that he and his family continue to attend proceedings as a matter of heart: "a pained heart," as he specified. Not only did they lose all of their sheep and their traditional livelihood in 2008, they also lost his wife Rosa to breast cancer in 2003.

The Masseria Carmine dates back to 1859, twelve years before Italian Unification was complete, when the marquise Filippo di Beaumont Bonelli of Naples purchased sixty hectares of land, upon which he established multiple farms. Research suggests that, in need of someone to help manage operations, the marquise (or more likely a son) hired Don Angelo's father, Vincenzo, sometime around the turn of the century. Fifty-odd years later the Beaumont Bonelli family decided to sell off all of the farm properties, and Vincenzo bought that which was closest to his heart, the Masseria Carmine, as it was there that he and his family had sought refuge during the bombings of World War II. A decade later the Fornaro family sold a small portion of that same land to the Italian government, for the construction of what was then called the Italsider steelworks. On the rest of it, they cultivated olives and grain, and raised sheep and other animals. Everything changed in March 2008, however, when the Asl (Azienda sanitaria locale, Local Health Agency) began a series of analyses of agricultural products hailing from within a twenty-kilometer radius of what had by then become Ilva. Eventually, the bodies of both goats and sheep from the Masseria Carmine revealed dioxin levels at three times the legal limit, and in September the order was given to slaughter a total of 1,200 feed animals in the zone, half of which belonged to the Fornaro family. During that same period, Ilva repossessed 2,500 olive trees that the Fornaros had cared for at the edge of the steelworks' land, and further analyses confirmed the high toxicity of the Masseria Carmine's soil.[46]

Since then, the Fornaro family has turned, with the aid of the group CanaPuglia and other regional environmental researchers, to the cultivation of hemp on the family's land. Last I checked, they had not yet been approved to sell the resulting plants for fiber and seed on the commercial market, but CanaPuglia and others were actively conducting analyses on the plants with the hope that this might change soon. In the meantime, the Fornaro family—led by Angelo's son Vincenzo, their most outspoken advocate—regularly invites both community groups and media outlets to the Masseria Carmine so that others might learn of hemp's ability to pull toxins out of the soil. Whether with visiting reporters or groups of area children, Vincenzo and family invite visitors to walk the fields together, all the while discussing hemp's potential to provide raw material for goods, the production of which could reinvigorate art forms once common in the region and ultimately provide economic opportunity for area residents. In 2014, the family hosted a public event in which they welcomed the local community to join them in planting that year's hemp crop. The event is described on the Masseria Carmine's website as "the umpteenth attempt at resistance and to show that our splendid land can continue to live!!"[47]

As Vincenzo Fornaro details in the documentary film *Non perdono*, discussed in chapter 5, hemp offers Taranto multiple forms of recovery: of physical health for land and future inhabitants, of local labor opportunity, and of artisanal practice—all after having grown in and through contaminated soil. He is steadfast in delivering his message, not only in the film but also in a growing number of news articles. In the photographs that tend to accompany such articles, he consistently appears posed within strikingly hearty fields of hemp sprouting waist-high, blurring the image of steelworks off in the distance.[48] As I discuss further below, I read Vincenzo Fornaro's work—inviting others to join him and his family in their fields, giving interviews to media, and, in 2017, running a political campaign to serve as Taranto's next mayor—as overtly performative: a stylized repetition of acts that work toward shaping a reality.

Also performative is the work of Noel Gazzano. Trained as a sociocultural anthropologist in North America, Gazzano has worked as a visual and performance artist in Italy for more than a decade. As her website explains, her work is dedicated to "promoting social transformation regarding gender issues and the current ecological crisis."[49] Applying ethnographic research methods toward the production of print, performance, and installation pieces, Gazzano is particularly attuned to themes of gender-based violence and community response to extreme pollution. Uniting the two, her work speaks to some of the same fundamental questions present in Seveso—not so much about abortion but about the broader correlations between environmental toxicity and women's reproductive bodies, as well as the community caretaking roles so often filled by women.

In April 2016 Gazzano undertook one of her more ambitious projects, *L'Insopportabile contraddizione—The Unbearable Contradiction,* in which she walked one hundred kilometers across Puglia, from Brindisi to Taranto, to draw attention to what she calls the "unbearable contradiction" between the region's physical beauty and the extreme health risks posed by its environmentally abusive industries. As she walked, she met with local residents, invited them to walk alongside her, slept in their homes and fields, all while pushing a hospital gurney sprouting hemp seeds, which she then planted upon arrival at her final destination, the Fornaro family's Masseria Carmine.[50] Her journey was heavily documented by various media outlets, and Gazzano herself later compiled a video record titled *Terra Mia, Io Cammino per Te* (Oh my land, I walk for you), which she has screened throughout Italy and North America. In the five-and-a-half-minute video, viewers see brief clips of Gazzano arriving in town centers throughout Puglia: old men in suits approach her to link arms in warm solidarity, children offer bouquets, and organized welcoming committees make speeches about her courage in the face of polluting monsters.

Primarily, though, the video offers a series of striking still images of Gazzano in the surrounding countryside, always wearing a full-length red cotton dress and often with the white hospital gurney held out in front of her waist. She is mostly shot in profile or from behind as she walks down roads lined with wildflowers and cactus, gazes out at olive groves (so many of which are afflicted, in Puglia, with the plant bacterium *Xylella fastidiosa*), or approaches Taranto's industrial cluster along the waterfront.[51] The images are accompanied by a poetic voice-over, an ode to the land in which Gazzano speaks of corporeal connections between feet and soil, sun and skin, toxins and breath. In a steady voice she declares: "I love you, my land . . . you can even say that I'm crazy in love, I'm yours. Your clumps of dirt are my flesh, your olive branches my skin, and the poison that they spread across you, my land, I feel it in my veins and it invades all the way to my mouth, where there were once only honey and almonds and the most bitter taste we knew was leftover chicory."[52] Much like the mother character in Argentina's *Vicolo dell'acciaio,* who breathes Taranto just as it breathes her,

Gazzano identifies as belonging to the broader Pugliese landscape just as it belongs, in a sense, to her: I am yours / you are me. In this, her work brings us back once more to a clear explanation of transcorporeality in ways both poetic and sobering.

Most visually striking in Gazzano's video, as well as in the additional still images and clips to be found online, is the presence of the aforementioned hospital gurney. When I first began speaking about Gazzano's work in conjunction with a larger Tarantine narrative movement, I presented a misreading of this particular element. Based on a casual conversation with Gazzano and a hasty viewing of early video footage, I had thought that the gurney was simply a bed and that Gazzano dropped the hemp seeds while she walked, from a hand held low and close to her body. In this (mis)reading, the project explored traditional gender roles for southern Italian women, who are so often tied to the physically generative potential of the female sexed body. By dropping seeds from near her own womb, I imagined that Gazzano was acknowledging cultural associations between women, reproduction, and caretaking, while in fact using the plant and its detoxifying potential to explore a futurity not tied to human reproduction but to the intransitive or "avuncular stewardship" of which Sarah Ensor has written.[53] It does not take a very focused second viewing, however, to reveal that the bed in question is most certainly a hospital gurney and that it serves as commentary on the connection between human and environmental health, as well as a site for materially transformative action.

As Gazzano herself writes on her artist website: "On the gurney I sprouted phytopurifying hemp seeds and planted them upon arrival to state that real cancer prevention consists in land-cleaning."[54] Drawing attention to the high rates of actual illness in Puglia, the hospital gurney simultaneously counters future illness by fostering redemptive and remediative vegetation for the polluted soil closest to the Ilva steelworks. My initial misreading was not entirely wrong: in her full dress and long, flowing hair Gazzano's self-presentation still incorporates traditional emblems of southern Italian femininity, and some of the most vocal environmental activists in Taranto and environs identify explicitly through their associations to motherhood, such as the group Mamme da Nord a Sud (Moms from the North to the South). The fact that the gurney is not simply any bed but the sort of bed used in hospital settings—to transport patients from one treatment to another or to support them as they move from the realm of the living to the dead—is significant, and its use in Gazzano's work as a site of slow, careful growth reads as an act of radical hope in the face of such loss. At the same time, it gives physical representation to the broad range of caretaking responsibilities so often held by women, but left unarticulated in policy discussions and dominant representational norms.

Through their individual land-based practices, Fornaro, Gazzano, and their community collaborators position hemp as dioxins' opposite. It is purifying rather than toxic, and so manifestly perceptible in its overt and multiform physicality as first seed, then plant, then eventually fiber and beyond. In both reality and representation, Taranto's hemp and its dioxins thus exist in a particular sort of eco-dialectic, each requiring material terrain— soil—to physically hold, facilitate, and filter disparate experiences of transmutability and transcorporeality, of being and becoming. What's more, both substances, especially with the aid of Fornaro and Gazzano's actions, force interlocutors to really notice that terrain, to think about the land spreading beyond the Ilva factory, and the many lives that land has led—as well as the lives of the many other beings that it can still affect.

To return to those collective guiding terms identified at the start of this discussion, the work of Fornaro, Gazzano, and others working in this vein is steadfastly one of resistance (to large industry and its offenses) and recovery (of land and bodies). While their projects differ in key ways—Gazzano's a conscious work of art, Fornaro's one of community action—I maintain that they both also engage aspects of theatrical performance. What's more, the performative and participatory natures of their projects are crucial to a deeper cognitive and emotional engagement with the primary matter at stake, Taranto's troubled landscape, as well as the dialectical pair of substances moving through it. While Gazzano, draped in a striking red, invites community members to walk with her as she prepares to plant seeds from a hospital gurney, Fornaro and family regularly welcome media outlets to document their crops' growth on the immediate border of Ilva's territory, just as they have welcomed collaboration from Gazzano, the artists of *Non perdono*, and members of the wider community. In describing their land-based actions as performative, I am of course thinking of Eugenio Turri's aforementioned theory of "landscape as theater" but also of the work of the contemporary performance studies scholar Erika Fisher-Lichte, who writes that theater "comes into being through the bodily co-presence of actors and spectators, through their encounter and interaction."[55] Readers might consider again the description of Gazzano's journey, punctuated as it was by her encounters and interactions with area residents as she passed through town centers, drawing them into her performance piece but also, ideally, into awareness and action.

The practices outlined above are marked by a copresence of people but also a copresence of human actants with the more-than-human world that emerges through embodied awareness. As Àlvaro Ivàn Hernàndez Rodrìguez writes: "We walk and become attentive to the unfolding world becoming ahead of us. We do not know what will be later in our walking as we do not know what was left behind, every step transforms into a new relationship and the traces we left are no longer the traces we are now. . . . Walking leads us to connection and collaboration—we never walk alone. Entangled occurrences woven together across time(s) emerge in every step, one and another, one after the other, one with the other."[56]

Walking or moving with assistive devices, such as wheelchairs, brings us constantly into a new present, just as we feel the many layers tucked into the land beneath us. We notice the presence of other elements or beings in our continually changing sphere of oxygen, just as we notice our own bodies adapting to movement. Joseph Dumit and Kevin O'Conner note, too: "We are made by use. We can be remade by different use. Our abilities to move . . . are the ongoing result of how we have moved in the past."[57] The same can, of course, be said of the landscape—which, as Fornaro, Gazzano, and others remind us, is so like and, indeed, so connected to our bodies, just as it is so much more. The walking central to their work confirms this connection, using embodiment to underscore not just our enmeshment with the soil beneath our feet but the mutability of other forms of dynamic matter moving within it. In bringing coparticipants into the Tarantine landscape to witness the planting of hemp and acknowledge the presence of dioxins, they draw awareness to the constant becoming inherent in both substances, one that has so damaged the Tarantine soil and one that might just resuscitate it.

There is one other set of Tarantine landscape practitioners that I must mention in these pages, a group whose work exists in clear dialogue with that of Fornaro and Gazzano and embraces those aforementioned key terms "performing resisting landscape." In the shadow of the Ilva steelworks, just a few quick steps from Taranto's main port and train station, a group of women artists engages in a collaborative practice that stands in direct opposition to destructive extraction, toxic emission, and industrial speed. Formed in 2014, this collective is known as Ammostro. Their name comes from an affirmative expression in local slang that essentially means "cool," while also winking at the local habit of referring to the steelworks simply as "the monster" (an attribution of animacy heavily featured in Cosimo Argentina's *Vicolo dell'acciaio*, as discussed in chapter 4).

Although individually specialized in a variety of trades, including graphic design, illustration, leather tooling, and dressmaking, the group's shared focus is on what they call "natural silk-screening." In their workshop bordering the Tamburi district, that closest to the steelworks and housing the population most affected by illness in Taranto, the artists of Ammostro print their designs on decorative and utilitarian goods, including wall hangings, clothing, and shopping bags. They produce their own dyes from native plants, such as Scotch broom (*Cytosis scoparius*) and use fabrics crafted from hemp. Their final products often feature images of native creatures, like the bioremediative mollusks once abundant in area waters, in a direct call to viewers—whether the individuals who buy their artwork or those who see it on the moving body or living room wall of a neighbor—to remember Taranto's native creatures and natural resources, who are now suffering.

In the summer of 2018 I sat with the members of Ammostro at their newly opened workshop to learn more about their origin and ethos. In a conversation quoted more fully below, they explained: "Working with the products of the land, using the old methods, the memories of the land, we sought out a new direction, no longer using the usual colors for silk-screening but extracting our colors from nature, from indigenous plants, and so we started to do research and experiment. . . . Being in Taranto, obviously, in a moment in which a critical consciousness was also forming, we couldn't help but to ask what kind of impact our work had on the environment, so this was a direct consequence."[58] Of their collective philosophy, and of the ways in which their work contributes to an ever-growing counternarrative about and from Taranto, they said: "Everything that comes out—it seems rhetorical, but it has a story to tell. . . . We try to identify signs and symbols of the land that recount something about Taranto, but also serve as vehicles or messages of resistance. For example, we did a whole series on the living creatures of the Mar Piccolo based on photographs from a marine biologist affiliated with Bocconi University."

The artists of Ammostro also spoke at length that afternoon about a form that appears frequently in their work: a small woven basket that was once commonly used by area fishermen to gather mussels. They craft miniature replicas of the basket from hemp fiber, to be used as decorative but perhaps also functional household objects, and they print its images on reusable shopping bags with ink from the aforementioned Scotch broom plant. Reading these basket pieces as narrative objects, as they suggest we do in the conversation cited above, allows for multiple stories to emerge. To begin, their material composition models an engagement with the surrounding nonhuman landscape that stands in stark opposition to that practiced by the nearby steelworks, oil refinery, and more. Their fibers and dyes are produced not through boring deep holes into the land only to blow and burn toxins out into the air but rather from plucking what emerges just above the soil—plants that are either native to the region or actively working to purify it—then carefully pulling the colors and crafting the material that these plants offer, without adulteration or toxic chemical manipulation.

Once in their final form, Ammostro's products avoid images of the Ilva steelworks or its name, which otherwise visually dominate both the landscape in and discourse about Taranto. Instead, they feature symbols, like baskets and mussels, from Taranto's traditional forms of livelihood, which were based on once-abundant natural resources. By recalling and re-creating in decorative form tools and practices that are no longer used, Ammostro's pieces tell of Taranto's history but also of its changing lifeways, gesturing toward the region's relatively recent transition from agricultural and maritime-based economies to an industrial one. At the same time, they remind us that there are alternative ways of engaging with the nonhuman landscape—if no longer through gathering mussels, then through making eco-conscious art. In this they contribute to the larger current of counterhegemonic storytelling taking place in Taranto. To create and disseminate stories of Taranto's past and possible presents and futures, while acknowledging through absence the harm wrought by the steelworks on community and land, is to also dismantle the sorts of majoritarian, pervasive, and thus toxic narratives that "blame the exposed communities for their lifestyle or naturalize a capitalist made disaster as a tragic accident."[59]

The artists of Ammostro make their wares widely available, selling them out of their storefront, online, and at festivals. They also share their craft, both the methodology and the final results, through collaborative workshops. It is particularly in this way that the members of Ammostro perpetuate a counternarrative about Taranto. Collectively crafting objects of beauty and use made with local resources, they show that Taranto is a place of creative capable residents and a naturally rich landscape rather than one marked only by strife and factory dependence. Their workshops are always held in public spaces, whether outside in a neighborhood piazza or in the open courtyard of one of the Città Vecchia's magnificently crumbling palazzos. In such workshops they might, for example, teach the children of the Tamburi neighborhood how to turn spoiled fruit into stamps for art prints to hang on their walls. Much like Fornaro and Gazzano, the Ammostro artists thus engage in a practice that is both pedagogical and performative. In this they bring what Doris Sommer calls "acupunctural art" to Taranto, not just sharing their creations but also encouraging others (both the children and passersby) to create their own art and, in the process, experience their own agency. Of responding to a work of art, Sommer writes: "Either you appreciate the stimulation of emotional and mental facilities as a satisfying re-enchantment of the world and stop there (in the tradition of art for art's sake); or your feeling kindles curiosity and arouses energy for making more art."[60] With interactive workshops, Ammostro invites participants to move directly from appreciation into action, and thus into their own process of creation, recognizing that "the very activity of art-making develops skills and imagination; it wrests some creative control over material and social constraints that might otherwise seem paralyzing."[61]

As a small local arts collective that makes both decorative and utilitarian objects, Ammostro is not unlike collectives we might find in other international cities such as Detroit, to which Taranto is occasionally compared. This is to say, Ammostro is not alone in the kind of work that its members are doing. Furthermore, the work of Ammostro alone, as lovely as it is, will not shut down the Ilva steelworks. And yet, through their humble practice, one that is manual, collective, and intentionally dictated by what is available in their dynamic more-than-human environment, they model a form of active resistance, while also nourishing multiple forms of regeneration. In offering a counterexample to nearby extractive, combustive, and heavily polluting industrial production, they tell a story of quiet resistance to such practices, while also modeling what Past describes as "slowness . . . as an ideology that can respond to shifts in environmental processes" and what they themselves call "a meditative slowness, the slowness that helps you to grow as a person."[62] By working deliberately within the natural limits of the nonhuman landscape, the artists of Ammostro are able to read its fluctuations, and thus to respect them. As an example, they spoke to me of searching for a plot of land where they could cultivate the plants they use for dyes, so as to not overtax precious resources in their landscape.

Engaging slowness as a rejection of extractive industry and a reclamation of the artists' own productive agency, the work of Ammostro stands as an ecological model inevitably tied, in factory-town-Taranto, to an economic one. It troubles assumptions regarding just who can afford the *privilege* of slowing down in what Franco Cassano has called our turbocapitalist economy, while exploring both the benefits and struggles of doing so. As such, theirs is also an exploration of value writ large. These women are highly skilled artists, with advanced degrees or specialized training from institutions located far from Taranto. They have chosen to either relocate or return to the polluted city, so as to make their eco-conscious work and contribute to the formation of a locally based counternarrative. They do not earn a great deal of money with this work (indeed, most of the artists have additional forms of employment), but they do produce meaningful connections to the people and land surrounding them, a privileging of affective experience over capital that can appear radical in the current moment. As Elizabeth Povinelli explains, "neoliberalism works by colonizing the field of value—reducing all social values to one market value—exhausting alternative social projects by denying them sustenance."[63] The work of Ammostro pushes back against that sort of neoliberal project, by actively and intentionally claiming both space and sustenance for their work, the sort of social project that, as Povinelli writes elsewhere, "is dependent on a host of interlocking concepts, materials, and forces that include human and nonhuman agencies and organisms," and is rooted in understanding of our deep enmeshment with the more-than-human world.[64]

Landscape is made, and makes itself, by use—of plants, animals, dioxins, humans. Like our bodies, landscape too is the result of how it has moved in the past, as well as how it is moving, and what is moving through it, in the present. Together, the work of Fornaro, Gazzano, and Ammostro builds a story of human actors performing resistance via landscape, perhaps even of human actors resisting destruction alongside landscape, through collective performative acts. They also draw our attention, however, to a story of the landscape performing a resistance of its own, working at root level—perhaps just backstage, or in fact below it—to undo some of the damage that we have caused. The examples outlined above represent just some of the creative work taking place in and about contemporary

Taranto. They are in deep dialogue with more traditional forms of activism and advocacy and have been bolstered in recent years by institutional support from the city and broader Puglia region for large-scale cultural events, such as the MEDIMEX International Festival and Music Conference. As I write, ArcelorMittal is actively seeking to withdraw its ownership of the Ilva steelworks, and a renewed energy crackles through the voices in Taranto that say it is only a matter of time until complete closure is achieved. In the meantime, the blast furnaces are still operating, illness is still rampant, and additional stories are no doubt ready to be shared.

FIRST PERSON

The Ammostro Artist Collective

After having followed the work of the Ammostro artist collective online since my initial visit to Taranto in 2016, I met with them in their brand-new workshop in the summer of 2018. Located in a sprawling old warehouse near the train station, the front-facing boutique portion they had planned for the space was not yet open to the public, but the artists had already begun to produce materials in their open workspace in back. Daniela Spera accompanied me on the visit, as we had spent the day together at the "Ambiente Svenduto" Ilva trial. We arrived already exhausted from a day of sitting in court but were reinvigorated by our lively exchange with Ammostro. Perhaps due to that initial exhaustion, I neglected to signal who was speaking when in my audio recording of the conversation. As such, I present the transcript here without speakers' names attached, so as to avoid false attribution of any statements. My hope is that this format might help to further underscore the collective nature of Ammostro's work.

> MONICA: Where does the name Ammostro come from?
> [*Laughter*]
> AMMOSTRO: Well, it's an expression . . .
>   Slang.
>   Local dialect, let's say.
>   . . . that means cool, great, *fico*. "Ammostro!" It's actually used a lot in Naples too. We've discovered that younger kids don't use it anymore. None of us are actually as young as we look. [*laughter*]
> MONICA: But where does it come from, what word?
> AMMOSTRO: From the word "monster"—and for various reasons. The monster, with the monstrous face.

Something so over-the-top to be almost monstrous, you know? Monstrously beautiful.

Yes, monstrously beautiful.

The [monster] mask, which has become the logo for our project, is from a *rosta* that you can see on Palazzo De Bellis in the Città Vecchia.[1] These masks, it seems, were important and positioned near entrances, hence the antefixes and dripstones, as a sign of protection—the "protective monster." And then of course in Taranto there is another famous monster that, instead, kills. So we wanted to play, to use this word to contrast a monster that protects with a monster that kills.

MONICA: Do you have a "mission statement," a shared vision?

AMMOSTRO: Yes, you can say that everything that comes about. . . . it seems rhetorical to say this, but it has a story to be told. More specifically, we try to identify signs and symbols from the territory that recount something of Taranto but also convey a message of resistance. For example, we did this whole project on all the living creatures of the Mar Piccolo based on photographs by a marine biologist, Rossella Baldacconi. And from there we developed a series of prints that we put on T-shirts, cloth bags . . . and then, we did this other project about the *nassa* [fish or crab trap basket]. What you see is actually a new interpretation. The *nassa* is a fishing instrument with a bell inside that imprisons the fish. We don't like this idea of prison or death, and so we got rid of that part. We asked an artisan in Massara to make us one like this [shows me a small basket], removing the internal bell, and we started to use it to package our products, and then we also made graphics inspired by the *nassa*.

Here is a trial screen-print, made with natural dyes from Scotch broom and *Reseda* plants. With this project, working with the products of the land, using the old methods, the memories of the land, we sought out a new direction, no longer using the usual colors for silk-screening but extracting our colors from nature, from indigenous plants, and so we started to do research. And from this research we started to experiment, like this [*gesturing toward the print*]. It's from a local plant, a Mediterranean plant, from which we get this yellow. So anyways we did a trial screenprint, with the same image of a *nassa,* seen from above.

Being in Taranto, obviously, in a moment in which a critical consciousness was also forming, we couldn't help but to ask what kind of impact our work had on the environment, so this was a direct consequence of the research that we started to substitute and make all of our dyes natural. And we use all vegetable resin, too, because we've also made this choice (none of us eat meat). I think, we think, that we share all these things and then the work that we do is also very much in line with the personal evolution that we've all gone through.

MONICA: Where do you see yourselves in two years, five years?

AMMOSTRO: We won't leave this place that has invested in us!

I can see something that would be in line with this whole "nature project": a green space where we can dedicate ourselves also to the cultivation of dyeing plants, because it's true that we use plants that grow wild, but it's also true that some of them are protected species. So it makes sense if we're going to produce on a larger scale to cultivate plants, no? For example, one of the inks that we've used a lot comes from myrtle berries. . . . These questions are slowly emerging.

Also, regarding the choice of fabrics, for example: in the beginning we obviously didn't have a lot of choice because we didn't have the economic resources to right away make a certain kind of investment. But then, moving from standard fabrics at good prices, we decided to choose certified fabrics, like those made from hemp or nettle.

MONICA: And Taranto, what will it be like in five years? Do you see a Taranto without Ilva?

AMMOSTRO: That's what we wish for, but in the future. Honestly, I don't think the situation will be that different in five years. That is, it's so serious it will take longer than that! . . . Five years from now . . . if the day comes, I don't think that's enough time.

This is a hard question.

It's hard.

It's a strange paradox. Maria is from Tamburi and she's shown us so many crazy places, abandoned places, and you say "shit, let's make a project here, let's try to find some funding, create connections, get involved in this way." But then you say, "shit, if they don't close Ilva what's the point?"

We should say, though, that it's not just Ilva. The mentality of the average *Tarantino* also needs to change a little bit. Because if these places remain abandoned and continue to be vandalized, treated like open-air dumps, there's not a lot we can do.

If the average citizen doesn't become more aware, the same things will always happen, the decisions will continue to be made on high and we'll just have to accept them.

It's self-determination that can change things, everyone doing what they can.

It's a sort of bond, right? I think that everyone should do a little something toward a shared goal.

[As an example] we ran this workshop about salvaging fruit. Every Wednesday in the Salinella neighborhood they have this market, and [we sent] children to go to the fruit and vegetable stands to salvage the damaged fruit, the fruit that was maybe a little old.... And then we had a healthy snack break during the workshop days, fruit salads to die for, with strawberries and cherries, and we [also] made dyes.

And we had positive feedback: after the course the kids told us, "Do you know that we're going to the market every Wednesday for fruit? We're having a healthy snack break." So with this we planted a little seed.

MONICA: What would you want a non-Italian audience to know about Taranto? What are the important points that I should include in my book?

AMMOSTRO: For me, definitely, that it's not just Ilva. That luckily there are those who resist and want to fight.

And also all the beauty here. For me, personally, it's limiting to just talk about Ilva.

That it is definitely not a land of industrial vocation....

Far from it!

That the potential is here, but not the desire from above to let this potential blossom.

And also that it's just an incredible place, from the climate to the natural beauty here. It's impossible for it not to be a source of inspiration. There is so much potential to live well in Taranto.

June 12, 2018

Notes

INTRODUCTION

1. Kington, "Italian Town Fighting for Its Life." As of this writing, Ilva's former leadership is still embroiled in a massive legal battle, in which prosecutors claim the company is responsible for 11,550 premature deaths in a seven-year timespan.
2. Nixon, *Slow Violence*.
3. Adam, *Timescapes of Modernity*, 54–55.
4. Buell, "Toxic Discourse," 665.
5. Armiero, "Introduzione," 16.
6. Ibid.
7. Nixon, *Slow Violence*, 15.
8. The European Agency for Safety and Health at Work, for example, requires EU member states to establish national occupational exposure limit values based on an ever-changing list of substances and figures.
9. Liboiron, Tironi, and Calvillo, "Toxic Politics," 334.
10. Ibid., 333.
11. Bianco et al., "Dibenzo-*p*-Dioxins and Dibenzofurans in Human Breast Milk."
12. U.S. Government, *The Health Risks of Dioxin*, 3.
13. Environmental Working Group, "Dioxin Timeline."
14. Beck, "The Love Canal Tragedy."
15. As noted in Fetters, "The Tampon: A History."
16. Yoshida et al., "Japan's Waste Management Policies."
17. World Health Organization (WHO), "Dioxins and Their Effects on Human Health."
18. Agency for Toxic Substances and Disease Registry (ATSDR), "Toxicological Profile."
19. Ibid.
20. Environmental Working Group, "Dioxin Timeline."
21. Agency for Toxic Substances and Disease Registry, "Public Health Statement Chlorinated Dibenzo-p-Dioxins (CDDs)," 10.
22. Alaimo, *Bodily Natures*, 2.
23. Bennett, *Vibrant Matter*, 122.
24. Chen, *Animacies*, 11.

25. Alaimo and Heckman, *Material Feminisms*, 4.
26. For more, see https://hmpdacc.org/hmp/.
27. Pollan, "Some of My Best Friends Are Germs."
28. Cohen, *Stone*, 3.
29. Dumit, *Drugs for Life*.
30. Appadurai, *Modernity at Large*, 8.
31. Iovino, "Toxic Epiphanies," 50.
32. See https://www.collinsdictionary.com/dictionary/english-thesaurus/receptive.
33. Herman, "Narratology as a Cognitive Science."
34. James, *The Storyworld Accord*, xv.
35. Heise, "Eco-narrative."
36. Donly, "Toward the Eco-Narrative," 6.
37. Knickerbocker, *Ecopoetics*, 159.
38. Caracciolo, *Strange Narrators in Contemporary Fiction*.
39. Verdicchio, *Ecocritical Approaches to Italian Culture and Literature*, xv.
40. Cesaretti, *Elemental Narratives*, 5; Iovino, *Ecocriticism and Italy*, 9.

1. SEVESO

1. Fratter, *Seveso*, 23
2. Bertazzi et al., "The Seveso Studies," 625.
3. Ibid.
4. Thomas and Spiro, "An Estimation," 1.
5. Laura Centemeri notes that in 1976 it was not yet possible to measure the level of dioxins in blood, but only in soil (Centemeri, *Ritorno a Seveso*, 26). Needham et al., "Serum Dioxin Levels," 225. Additionally, as of June 2014 the World Health Organization notes: "The quantitative chemical analysis of dioxins requires sophisticated methods that are available only in a limited number of laboratories around the world. The analysis costs are very high and vary according to the type of sample, but range from over US$ 1000 for the analysis of a single biological sample to several thousand US dollars for the comprehensive assessment of release from a waste incinerator" (WHO).
6. Rocca, *I giorni della diossina*, 25.
7. "La mancanza di chiarezza e i contrasti che interagirono fra i diversi Enti furono percepiti chiaramente dalla popolazione che non riuscì, nel momento dell'emergenza, a trovare nell'autorità, con tempestività e chiarezza, quelle risposte di cui avrebbe avuto bisogno di fronte ad un veleno 'invisibile'" (Fratter, *Seveso*, 111). All translations from the Italian are my own unless otherwise indicated.
8. Accessed May 28, 2015.
9. "La dipendenza dalle informazioni fornite dalla Givaudan rendeva difficile alle autorità locali decidere la linea da seguire" (Centemeri, *Ritorno a Seveso*, 26).
10. (1) è estremamente importante lavare immediatamente ed a lungo le mani.... Si consiglia anche una frequente ed accurata pulizia di tutto il corpo (bagno

o doccia tutti i giorni). . . . (2) essendo noto che i soggetti venuti a contatto con sostanza possono andare incontro a manifestazioni cutanee da ipersensibilità alla luce solare, si consiglia di non esporsi al sole per periodi prolungati. (3) l'ingestione di alimenti contaminati provoca una diffusione della sostanza nell'organismo con danni prevalentemente localizzati al fegato e al rene. Perciò è altamente pericolosa l'ingestione di qualunque alimento animale o vegetale prodotto nelle zone sopra indicate. È pertanto vietato coltivare o raccogliere nelle zone inquinate foraggio, erba, fiori, frutta, verdura, ortaggi, nonché allevare animali accetti quelli d'affezione. (4) è vietata ogni lavorazione che provochi movimento di terreno, in quanto comporta contatto cutaneo e sollevamento di polvere. . . . (5) al fine di evitare sollevamento di polvere, la velocità degli autoveicoli su strade non asfaltate non deve superare i 30 km orari. (6) è prudente che tutte le persone esposte al rischio della contaminazione si astengano dalla procreazione per un periodo di tempo che cautelativamente può essere indicato in almeno sei mesi. Accessed at the Fondazione Luigi Micheletti archives in Brescia, Italy, May 28–29, 2015.

11. "Una sensazione strana. Questa me la ricordo bene. Se uno dice 'Ho mal di stomaco, ho mal di testa' è una realtà che sai. Ma questa cosa invisibile. . . . E questa sensazione che mi sto portando dietro, nel senso . . . non che . . . Nel senso di qualcosa di non palpabile che ti può succedere da un momento all'altro. Allora eravamo terrorizzati" (Centemeri, *Ritorno a Seveso*, 35).

12. Biehl and Moran-Thomas, "Symptoms," 273.

13. "Il rischio 'invisible' della diossina, per essere reso percepibile e presente, avrebbe dovuto imporre comportamenti estremamente rigorosi, coerenti e realistici da parte di chi era guardato come detentore del sapere, un sapere che dalle persone, veniva dedotta dal fare" (Centemeri, *Ritorno a Seveso*, 45).

14. "l'incertezza sul fare veniva interpretata come incertezza del sapere, e quindi incertezza sulla pericolosità della diossina" (Conti, *Visto da Seveso*, 25).

15. "Non ho più visto rondini e quando non si vedono più rondini è brutta, perché è veramente successo qualcosa e quando è venuta fuori la diossina di rondini non se ne sono viste più, sparite tutte" (Fratter, *Seveso*, 21).

16. The children who bore the most acute signs of chloracne continued to suffer for decades to come. A 2015 article in the Milan edition of the *Corriere della Sera* chronicles the trials of those affected, as neighbors blamed them for profiting finanancially from their scars: http://milano.corriere.it/notizie/cronaca/15_agosto_27/noi-bambini-diossina-ancora-trattati-come-colpevoli-afb598b0-4c82-11e5-9b47-ed94dd84ed07.shtml?refresh_ce-cp.

17. Consonni et al., "Mortality in a Population Exposed to Dioxin."

18. "La nostra ipotesi è allora che alle spalle della stampa abbiano agito gruppi di interesse e di pressione che si sono adoperati per strumentalizzare il caso di Seveso, trasformandolo in un'occassione propizia per colpire avversari . . . e, nello stesso tempo, rafforzare la propria ideologia e il proprio potere sull'opinione pubblica . . . le vittime di questo gioco senza scrupoli" (Mascherpa, *La stampa quotidiana*, 82–83).

19. See http://ec.europa.eu/environment/seveso/legislation.htm.

2. SEVESO STORIES

1. Barca, "Lavoro, corpo, ambiente," 542.
2. Full titles can be found in Works Cited.
3. Barca, "Lavoro, corpo, ambiente," 550.
4. "Studio ecologia perchè sono curiosa di sapere come funziona il meccanismo della vita . . . amo gli animali, amo la specie umana e vorrei che durasse a lungo" (Boldrini, "Intervista," 1991).
5. Barca, "Work, Bodies, Militancy," 121.
6. "Tutto quello che potevamo dire era che coloro che vivevano sulla terra inquinata avevano 'una certa probabilità' di assumere nell'organismo una sostanza che avrebbe 'aumentato la probabilità' di ammalarsi. Questa 'probabilità al quadrato' era difficilissimo non tanto da spiegare intellettualmente, quanto da fare entrare nelle emozioni, così che suscitasse l'emozione del timore e quindi il desiderio di allontanarsi" (Conti, *Visto da Seveso*, 55).
7. *The Economist*, 1977 (n.p.), photocopy accessed at the Fondazione Luigi Micheletti archives in Brescia, Italy, May 28, 2015.
8. "Nonostante il lancio pubblicitario e le recensioni forzatamente favorevoli, il libro non ebbe successo . . . un insuccesso meritato," *Il Corriere di Seveso*, n.p., photocopy accessed at the Fondazione Luigi Micheletti archives in Brescia, Italy, May 28, 2015.
9. "Cominciavo a rendermi conto che 'ambiente' non è solo l'insieme di acqua, aria, terra; che non si può considerare l'uomo nel suo rapporto con la natura se non lo si considera anche nel suo rapporto con gli altri uomini, e nel suo rapporto con gli oggetti che fabbrica o con le piante che coltiva" (Conti, *Visto da Seveso*, 85).
10. Barad, *Meeting the Universe Halfway*, 33.
11. Iovino and Oppermann, *Material Ecocriticism*, 1.
12. "Quando muore la materia che forma il suo corpo non resta ferma. Sarà mangiata da altri organismi, che a loro volta si muovono" (Conti, *Visto da Seveso*, 91).
13. "Vedevo, nell'immaginazione, il suolo come un immenso formicolìo di organismi in movimento, di sostanze in trasformazione continua" (ibid., 91).
14. "non ora, ma fra tre mesi, forse fra un anno, forse fra cinque anni, forse fra dieci anni, verrà una malattia; di fronte a queste mie asserzioni ho visto gente che rideva di cuore. Il futuro non esiste. Un vecchio, tenendo la mano sul capo della nipotina, mi disse: 'Cara dottoressa, anche questa creatura un giorno morirà. Ma siccome io non so *quando*, per me è come se non dovesse morire *mai*'" (ibid., 55).
15. Adam, *Timescapes of Modernity*, 54–55.
16. Ibid., 194.
17. "da libro di divulgazione, cioè educativo, diventava un libro su una particolare crisi del processo educativo. Quindi, da libro per ragazzi, diventava libro per adulti" (Conti, *Una lepre con la faccia di bambina*, 10).
18. "andavo accorgendomi, giorno per giorno, del problema della communicazione" (Conti, *Visto da Seveso*, 117).
19. Iovino, "Toxic Epiphanies," 42.

20. Lehtimäki, "Natural Environments," 127.
21. Again, see Iovino for more on "the inertia of the authorities, the missed integration and unequal protection, the social landscapes of marginality, thwarted citizenship, and the discriminating practices carried out on women's bodies," following the Seveso disaster (Iovino, "Toxic Epiphanies," 43).
22. Herman, "Narratology as a Cognitive Science," 9.
23. Scholes, "Language, Narrative, and Anti-Narrative," 205.
24. Conti, *Una lepre con la facia di bambina*, 64.
25. Ibid., 22.
26. And not, as Laura di Bianco reminds me, "A Child with the Face of a Hare."
27. Conti, *Una lepre con la facia di bambina*, 76.
28. Emily Dickinson, "Tell All the Truth but Tell It Slant," May 16, 2021, https://www.edickinson.org/editions/2/image_sets/12171689.
29. Conti, *Una lepre con la faccia di bambina*, 40.
30. Nussbaum, *Cultivating Humanity*, 90.
31. Ibid., 91.
32. Conti, *Una lepre con la faccia di bambina*, 77.
33. Iovino, "Toxic Epiphanies," 46.
34. Conti, *Una lepre con la faccia di bambina*, 58.
35. Ayuero and Swistun, "The Social Production of Toxic Uncertainty," 366.
36. Conti, *Una lepre con la faccia di bambina*, 105–6.
37. Ibid., 114.
38. Chemical toxicity such as that conveyed by dioxin exposure is of course only universalizing to a limited extent. As is repeatedly the case with toxic environmental disaster, the 725 residents of Seveso's Zone A were largely of lower socioeconomic means than other residents in the greater community. Like the family of Conti's fictional Sara, many of the 725 were factory employees and their dependents and had originally immigrated from Italy's impoverished south.
39. Conti, *Una lepre con la faccia di bambina*, 14.
40. Genette, *Narrative Discourse*, 161.
41. Iovino, "Toxic Epiphanies," 45.
42. Donly, "Toward the Eco-Narrative," 7.
43. Cited in James, *The Storyworld Accord*, 137.
44. Blanchot, *The Writing of the Disaster*, 60.
45. My attempts to find publication information were not successful. I am grateful to Angela Alioli and Gemma Beretta of Circolo Legambiente Laura Conti Seveso for their precious assistance in reaching Editori Riuniti, which is now under new ownership and does not have access to data for their older catalogue.
46. "questo libro mi è parso abbastanza interessante, ma a tratti noiosi" (letter from the Scuola media Canobbio-Lugano, classe 2G 1989/90, students of Nicola Canonica, dated March 30, 1990, and accessed at the Fondazione Micheletti, May 29, 2015).
47. Many in Seveso still felt offended by their representation on-screen. See, for example, the television program *Fluff, processo alla TV* from March 28, 1989. *Fluff,*

which ran for two seasons, was a talk show of sorts dedicated to the critical analysis of television as a communicative medium. The episode in question assesses Serra's adaptation of the film and includes a roundtable discussion between host Andrea Barbato, Serra, Conti, and two other guests I have been unable to identify, as well as a series of interviews with Seveso residents, most of whom express dismay at the film's out-of-touch representation of their community.

48. "Ma tra la minaccia della diossina e le reazioni, anche razziste, dei residenti di Seveso, si fa strada quasi poeticamente il modo di vivere il dramma da parte di Sara e di Marco" (Calcagno, "Due ragazzi," 1988).

49. "Delle molteplici chiavi di lettura della tragedia di Seveso, Gianni Sera ha privilegiato quella poetica, come se volesse sfuggire a quella politica. Ma *Una lepre con la faccia di bambina* è un film politico, con tutto ciò che ne consegue, anche se debole e talvolta latitante è la denuncia delle responsabilità" (Garambois, "Seveso fa ancora paura").

50. Although the director claims that he began to write the screenplay in 1978, "the day after the book came out," he sat on the project for ten years before completing it. As a 1988 article in *L'Unità* explains, "When Gianni Serra picked it back up this year, he instead found it 'current,' with the *Karin B.* navigating with its loads of poison from one port to another" (Garambois, "Seveso fa ancora paura," 1988).

51. See Steven Greenhouse, "Toxic Waste Boomerang: Ciao Italy!," *New York Times,* September 3, 1988.

52. "Se consideriamo che dopo Seveso, Chernobyl, sono di estrema attualità la Karin B e le altre navi con i loro rifiuti velenosi che si aggirano nel Mediterraneo, 'Una lepre con la faccia di bambina' assume un significato estremamente importante sia per i fatti di allora sia per quelli cui assistiamo in questi giorni" (Calcagno, "Due ragazzi," 1988).

53. Unlike in the book, the film's opening shot features ICMESA workers running from a smoking reactor on factory grounds, then transitions to Marco returning home from a camping trip and overhearing a radio broadcast describing the explosion.

54. Schoonover, "Documentaries without Documents?," 491.

55. Ibid., 505.

56. Cited in Past, "Trash Is Gold," 606. Original source: Vivian Sobchack, *Carnal Thoughts: Embodiment and Moving Image Culture* (Berkeley: University of California Press, 2004), 77.

57. James, *The Storyworld Accord,* 34.

3. TARANTO

1. It is important to underline that the Ilva steelworks is not the only culprit in Taranto when it comes to environmental health, but it is by far the largest.

2. Attino, *Generazione Ilva,* 53.

3. Ibid., 55; Giannì and Migliaccio, "Taranto, oltre la crisi," 159.

4. This was not the first such boom, although it was the most significant. Gianni and Migliaccio note that Taranto experienced an earlier haphazard construction boom at the start of the 1920s, when the naval industry provided many employment opportunities in the area.

5. "una città disastrata, una Manhattan del sottosviluppo e dell'abuso edilizio" (Cederna, "Taranto in balia dell'Italsider," 3).

6. Pasolini, *Il viaggio jonico*, 10, 7.

7. Attino reports that Taranto boasted the highest per-capita income growth in southern Italy in 1973 (Attino, *Generazione Ilva*, 54).

8. Servizi Parlamentari, "Camera dei Deputati—7-00514—Risoluzione presentata dall'On. Segoni (M5S) ed altri il 6 novembre 2014," 1–2.

9. Stefania Barca cautions that the decisions working-class communities make about their roles in environmental impact "are often overly simplified as 'jobs versus the environment,' which obscures the nature and diversity of environmental activism that develops from working-class ecological consciousness" (Barca, "Laboring the Earth," 14).

10. ARPA Puglia, "Cronologia emissioni da impianti agglomerazione ILVA 1994–2011."

11. Vagliasindi and Gerstetter, *The ILVA Industrial Site in Taranto*.

12. Barca and Leonardi, "Working Class Ecology."

13. Agenzia Nazionale Stampa Associata, "Ex Ilva: studio Sentieri, 600 bimbi malformati a Taranto," ANSA.it, June 1, 2019, http://www.ansa.it/canale_saluteebenessere/notizie/sanita/2019/06/01/ex-ilva-bonelli-verdi-nascosti-dati-su-600-bimbi-malformati_b966b4e1-30ce-4a1d-8731-b109c4fc6bb7.html.

14. Barca and Leonardi, "Working Class Ecology," 60.

15. ANSA, "Dioxins Found in Taranto Breast Milk."

16. Foschini, "Pecore tosiche abbattute."

17. Daniela Spera, interview by author, June 2016.

18. Ibid.

19. "la diossina entra nell'organismo e o non esce più o ci vogliono molti anni per smaltirsi" (Alessandro Marescotti, interview by author, June 2016).

20. See https://www.peacelink.it/ecologia/a/40532.html.

21. Alaimo, "Trans corporeality," 435.

22. Ibid.

23. Braidotti, "A Theoretical Framework for the Critical Posthumanities," 12.

24. Braidotti, *The Posthuman*, 82.

25. Hayward, "More Lessons from a Starfish," 68.

4. TOXIC TALES

1. As the sociologist Marianna D'Ovidio writes in her project on Taranto's notable graffiti culture, an immediate and broad-reaching form of narrative expression "se si va a Taranto non si vedono solo muri: si percepisce, quasi fisicamente, sulla

pelle, la presenza di un tessuto ricco di pratiche variegate che vanno dall'autocostruzione al cinema, dalla sartoria al teatro, dalle ceramiche al turismo sostenibile" (D'Ovidio, "A Taranto di muri ce ne sono tanti").

2. Vattimo, "Dialectics, Difference, Weak Thought," 50.

3. "questa maggiore attenzione al mondo esterno" (Maurizio Ferraris, "Il ritorno al pensiero forte," la Repubblica, August 8, 2011, https://ricerca.repubblica.it/repubblica/archivio/repubblica/2011/08/08/il-ritorno-al-pensiero-forte.html).

4. Mosca, "New Realisms or Return to Ethics?," 48.

5. Antonello, "Dispatches from Hell: *Gomorra* by Matteo Garrone," 380.

6. "la necessità di dire un vero che esorbita dai limiti dell'empiricamente accaduto" (Donnarumma, *Ipermodernità*, 126).

7. "lei negli anni aveva incontrato varie persone che le avevano raccontato le loro storie e lei stava facendo un dossier, perché la sua idea era appunto quella di fare ricorso alla corte europea, comunque stava raccogliendo delle storie. E io ho iniziato con lei a incontrare, ho fatto il suo stesso, lei l'ha fatto in 4 anni, io l'ho fatto in un anno, ad incontrare tutte queste persone che ci hanno raccontato le loro storie. Nel libro sono tutte persone vere" (Cristina Zagaria, interview by author, June 2016).

8. "quei pietosi casi simbolo" (Zagaria, *Veleno*, 69).

9. Certeau, *The Practice of Everyday Life*, 115.

10. Ibid., 155, 166.

11. ARPA Puglia, "Cronologia emissioni da impianti agglomerazione ILVA 1994–2011."

12. "penuria" and "pochezza" (Piccinni, *Adesso tienimi*, 35).

13. Cesaretti, *Elemental Narratives*, 128.

14. Simonetti, "I nuovi assetti della narrativa italiana (1996–2006)," 95.

15. "Sono nata a Taranto. 500 milioni di debiti e 90,3% della diossina che uccide l'Italia. Vivo in via Cagliari 32/A, in una villetta bianca con il cancello in ferro battuto arrugginito. Fumo due pacchetti di Chesterfield blu al giorno, mangio solo caramelle gommose senza zucchero e popcorn al formaggio. Nel tempo libero guardo la televisione o piango" (Piccinni, *Adesso tienimi*, 9).

16. "Il cielo a Taranto non è mai azzurro, anche quando ti sembra che sia così. Non dipende tanto dalla posizione geografica. Ma dall'intensità delle sfumature. Striate di rosso durante il giorno, di oro la notte" (ibid., 142).

17. "all'ILVA che colora il cielo e me lo fa sembrare più bello" (Piccinni, *Adesso tienimi*, 173).

18. Writing on the troubled protagonist in Michelangelo Antonioni's film *Red Desert* (1964), Past notes the character's "epidermic epiphany," explaining: "irreducible to her surroundings, Giuliana is yet of these surroundings" (Past, *Italian Ecocinema beyond the Human*, 47).

19. "la più alta percentuale di morti per cancro ai polmoni della penisola . . . il mare inghiottito dal mercurio, che il pesce lo sta drogando . . . i pomodori appesi a grappolo e le lenzuola, che si sono già colorati di rosso polvere. Rosso ILVA" (Piccinni, *Adesso tienimi*, 174).

20. Tuana, "Viscous Porosity," 198.

21. Zagaria, *Veleno*, 161–62.

22. "Sfiorano il cimitero, passano la rotonda ed eccoli sulla Statale 106. Ilva a destra, Eni a sinistra . . . Lasciano velocemente alle loro spalle la chiesa della Croce, Punta Rondinella, il molo polisettoriale, i container" (ibid., 137–38).

23. "Adriana mi mette la palma sulla scrivania. Si siede al bordo del letto e dice che stamattina, quando è uscita, non si respirava. Mi spiega che stava una specie di nube violacea che, le avevano raccontato, veniva dall'ILVA. Quando dice i veleni ci stanno entrando in casa annuisco. Vorrei dirle che è proprio vero, che i veleni ci stanno vicini e dentro, ma poi le mi sorride e indica la palma" (Piccini, *Adesso tienimi*, 76–77).

24. "una Taranto che è peculiare e universale" (Caprino, "*Adesso tienimi*, il libro d'esordio di Flavia Piccinni riletto dieci anni dopo").

25. "Più e meglio di tante analisi sociologiche e inchieste giornalistiche, in *Adesso tienimi* Flavia Piccinni racconta Taranto e i Tarantini" (ibid.).

26. Desiati, "Cosimo Argentina, Maschio adulto solitario," www.mannieditori.it/.

27. Prudenzano, "Arriva da Taranto e dal Salento la 'meglio gioventù' della nuova letteratura italiana."

28. "Via Calabria 75, un codominio dove il novanta per cento delle famiglie ha il capo che se la spassa nel siderurgico. Dal primo al settimo piano siamo tutti incrostazioni del grande tubo madre e la nostra pelle profuma del ferro zecchino delle acciaierie" (Argentina, *Vicolo dell'acciaio*, 9).

29. "Qui nel palazzo tutti muoiono di cancro ai polmoni, Il record della pista è nostro. Abbiamo in corpo, a famiglia, più benzene, polveri cancerogene, diossina, policarburi aromaticia e gas saturi di non so nemmeno io che cosa" (ibid., 12).

30. "Ogni volta che uno di noi esce dal quartiere è disorientato" (ibid., 163).

31. "I veri via Calabria al massimo vanno ad ascoltare un po' di musica dal vivo al Ramblas. Il Ramblas è un locale in zone Canale" (ibid., 53).

32. Ibid., 49, 18, 72.

33. Ibid., 62.

34. "è tossico pure a distanza" (ibid., 35).

35. "quasi come un antico nobile guerriero che forgia le proprie armi" (ibid., 176).

36. "Mio padre il Generale a quest'ora è lì, oltre i due ponti. Da dove mi trovo posso vedere la potenza del fuoco dell'Italsider. Posso vedere come il mostro gonfia i muscoli e si fa beffe di noi" (ibid., 175).

37. Chen, *Animacies*, 24.

38. Ibid., 30.

39. "L'Italsider occupa un appartamento di questo quartiere. Purtroppo 'st'appartamento è due volte Taranto. Dal suo domicilio, un centomila vani e accessori, il siderurgico se ne sbatte del divieto di immissioni, del divieto di atti di emulazione e di tutte le regole condominiali possibili e immaginabili" (Argentina, *Vicolo dell'acciaio*, 163).

40. As my colleague Stephen Sheehi reminds me, the word "corporation" (and Ilva is indeed a corporation) is derived from the Latin *corpus*. Throughout much of

the world, corporations are legally recognized as their own autonomous entity, much like a person. *Merriam-Webster Collegiate Dictionary* (11th ed.) tells us, in fact, that a corporation is "a body formed and authorized by law to act as a single person although constituted by one or more persons and legally endowed with various rights and duties including the capacity of succession." In this light, then, figuring Ilva as a giant bodily creature is not at all off the mark.

41. "Da lì prendo via Giovan Giovine ed emergo in via Dante . . . da corso Umberto finisco sul Canale . . . Attraverso il ponte e sono in Città Vecchia. La città araba la lecco sul fianco interno e poi sono nel rione Tamburi, e lì osservo il mostro, l'inferno d'acciaio" (Argentina, *Vicolo dell'acciaio*, 246–47).

42. "Via D'Aquino dà il senso dello stato in cui si trova la città. Una puttana a fine corsa, questo è diventata Taranto. Troia fino in fondo ma con i piedi dolenti e le caviglie a pezzi" (ibid., 183).

43. Iovino, *Ecocriticism and Italy*, 19.

44. "E invece lei è nel respiro di Taranto. È nel respiro pre-equatoriale. I suoi polmoni viaggiano sui binari Lecce-Battipaglia e hai voglia a ricordarle Alessandro Manzoni e la Madunina: non c'è storia . . . oramai è dei nostri" (Argentina, *Vicolo dell'acciaio*, 15).

45. "I fumi dall'Ilva entrano in cucina, in salotto, nel cesso. Aspiriamo diossina sotto forma di silenzi armati" (Argentina, *Vicolo dell'acciaio*, 127).

46. As well as the flow of so much more. Chen reminds us: "When physically co-present with others, I ingest them. . . . I am ingesting their exhaled air, their sloughed skin, and the skin of the tables, chairs, and carpet of our shared room" (Chen, *Animacies*, 209).

47. "Io rientro nella sala del chiavuto e mi riempio le narici dell'odore che Trottola comincia a emanare. Fizzo di ulcere infette . . . Il suo corpo sarà ormai un sacco di merda; lo è da almeno sei mesi. Dentro c'era solo questo alito, come un filo di bava di vita" (Argentina, *Vicolo dell'acciaio*, 73).

48. Armiero, "An Environmental Historian among Activists," 169.

49. As of 2015, approximately twelve thousand people were directly employed at Ilva's Taranto plant, with additional thousands employed as subcontractors. A report prepared for the European Parliament's Committee on Environment, Public Health and Food Safety from that same year raises concerns, however, about the economic impact of "the environmental problems caused by emissions from the plant," with cows and sheep repeatedly ordered slaughtered, the cultivation of various crops prohibited and tourism impeded (Vagliasindi and Gerstetter, *The ILVA Industrial Site in Taranto*, 7).

50. Colella, *Capatosta*, 21.

51. Foschini has also coauthored an investigative look at soccer, as well as a study of social habits related to fear in contemporary society.

52. "riunire interviste, rispettando l'alterità e la voce dei propri interlocutori" (Donnarumma, *Ipermodernità*, 80).

53. "la necessità di dire un vero che esorbita dai limiti dell'empiricamente accaduto" (ibid., 126).

54. "Crateri di cemento, magma e cassa integrazione, Taranto è un vulcano. Attivo" (Foschini, *Quindici passi*, 7).

55. "si trattava di un'allusione, che là dietro c'era qualcosa e che quel qualcosa fosse in qulache maniera legato alla città, al colore del cielo, al porto, alle ciminiere. In sostanza al Vulcano" (ibid., 15).

56. "in sostanza nell'aria di Taranto ci sono le sigarette" (ibid., 19).

57. Ibid., 21.

58. "la diossina non era soltanto nell'aria. O magari nella carne delle pecore o nei formaggi. La diossina era arrivata sin dentro le tette delle Tarantini" (ibid., 40).

59. "Io ho paura non soltanto di perdere il lavoro, io ho anche paura di ammalarmi e il discorso della malattia non è un discorso egoistico, ma si rispecchi in un passato e in un futuro oltre che nel presente. Comunque tieni conto che il peso di queste aziende così inquinanti lo ha patito chi c'era prima di noi, lo patirà quelli che arriverano dopo di noi" (ibid., 86).

60. Alaimo, *Bodily Natures*, 22.

5. ON-SCREEN ENGAGEMENTS

1. Hunter, "Nightmare Cities," 118.

2. The most reputable source I have found for this common claim again comes from Hunter, who writes: "Director Lenzi [has] gone on record as saying that the magnitude of the situation and its consequences inspired *Nightmare City*," but does not indicate to which record he refers (ibid.).

3. Ibid., 118–19.

4. Soles, "Sympathy for the Devil," 235. In his study of 1970s horror films, Soles cites Pat Brereton's work on Hollywood Utopia, in which Brereton "characterizes ecological thinking of the 1970s as concerned with 'conflicts embedded within humanity' as well as threats posed by 'technological developments and increased pollution,' concluding that this kind of ecologically focused 'paranoiac thinking' is reflected in the science-fiction and conspiracy thriller movies of the time" (234–35).

5. Jane Chapman offers the following succinct definition of the genre: "a discursive formation, presenting first-hand experience and fact by creating a rhetoric of immediacy and 'truth,' using photographic technology" (*Documentary in Practice*, 8), while Bill Nichols writes that "these films give tangible representation to aspects of the world we already inhabit and share. They make the stuff of social reality visible and audible in a distinctive way, according to the acts of selection and arrangement carried out by a filmmaker. They give a sense of what we understand reality itself to have been, of what it is now, or of what it may become" (Nichols, *Introduction to Documentary*, 1–2). Summarizing scholarly debate surrounding the definition of documentary, Kate Nash instead offers the following: "The failure to find a set of criteria sufficient to define documentary has been an ongoing theoretical issue within documentary studies," while emphasizing "the connection between ethics and aesthetic choice" (Nash, "Documentary-for-the-Other," 225, 227).

6. Nichols, *Introduction to Documentary*, 2, my emphasis.

7. See Nash, "Documentary-for-the-Other," for a wonderful overview of the evolution of observational documentary, based largely on Stella Bruzzi's *New Documentary* (2006).

8. Barca and Leonardi, "Working-Class Communities and Ecology," 66.

9. Caminati and Sassi, "Notes on the History of Italian Nonfiction Film," 361, 364.

10. Phelan, "Narrative as Rhetoric," 218.

11. O'Leary, "What Is Italian Cinema?" 11.

12. Angelone and Clò, "Other Visions," 85.

13. Ibid.

14. For more on the long history of Italian documentary, see Bertozzi, *Storia del documentario italiano;* and Ben-Ghiat, *Italian Fascism's Empire Cinema*, both listed in this volume's Works Cited.

15. Bondavalli, "Broadcasting Human-Animal Friendship: RAI Television's *I nostri amici* and Early Italian Ecodocumentaries," 191.

16. Seger, "Aesthetic Engagement: Anna Zanoli's Cultural Programs for the RAI (1968–1974)," 536.

17. MacDonald, "The Ecocinema Experience," 19.

18. See chapter 4 in Bozak's *The Cinematic Footprint*.

19. MacDonald, "The Ecocinema Experience," 20.

20. Willoquet-Maricondi, *Framing the World*, 45.

21. I cite again Erin James, here substituting "viewing" for James's original "reading" (James, *The Storyworld Accord*, xv).

22. Ingram, "The Aesthetics and Ethics of Eco-Film Criticism," 44.

23. Nichols, *Introduction to Documentary*, 116.

24. Fondazione Cineteca di Bologna, "Cecilia Mangini e Lino del Fra," http://www.cinetecadibologna.it/biblioteca/patrimonioarchivistico/fondo_mangini_delfra.

25. Missero, "Cecilia Mangini," 56.

26. Ibid., 55.

27. *Ferrhotel* won the UCCA Venti Città award at the 2011 Torino Film Festival.

28. "Non ero convinta di tornare sui luoghi dei miei film ma la storia di Taranto è la storia dell'Ilva, e prima dell'Italsider. Il confronto tra un'epoca che sembrava entrare nell'era moderna, quando giravo io, col petrolchimico e l'Ilva oggi diventava vitale" (Piccino, "Viaggio in Puglia").

29. "Il microcosmo che raccontiamo è per noi la chiave d'accesso a uno stato delle cose più ampio. C'è un messaggio che si ripete nel film: 'la gente deve reagire.' Ma anche: 'la politica non fa più il suo dovere'" (ibid.).

30. Past, *Italian Ecocinema*, 5.

31. "Siamo affogando, siamo completamente morti . . . dicono che hanno inquinato . . . poi dicono che c'è la diossina, no . . . noi non abbiamo capito niente."

32. Valeria Castelli further underscores that Barbenente's introductory address to viewers (revisited in a second voice-over toward film's end) "establishes her

'authority' as auteur, underlining her role as co-director of the documentary, and not only as Mangini's travel companion" (Castelli, "The Filmmaker Is Present," 241).

33. Ibid., 253.

34. Ibid., 241.

35. "È estate, e un giudice a Taranto ha ordinato il sequestro della più grande acciaeria Italia, e l'arresto del proprietario, Emilio Riva, per disastro ambientale, e il lavoro di molto operai è a rischio."

36. "voglio provare a guardare quello che sta avvenendo dal suo punto di vista."

37. Caracciolo, *Strange Narrators in Contemporary Fiction*, 29.

38. Missero, "Cecilia Mangini," 65.

39. Mangini, "Film Notes—Essere Donne," https://festival.ilcinemaritrovato.it/en/film/essere-donne/.

40. "Oggi l'industria si è affacciata anche tra gli ulivitti del mezzogiorno, ma lo sviluppo industriale del sud è lento, e insufficiente. Le raccoglietrici di ulivi sanno che la fabbrica è un passaggio obbligato per sottrarsi alle condizioni di lavoro arretrate, al rapporto patriarcale con la famiglia, e con l'uomo."

41. "ma le fabbriche sono poche. Ecco perchè dal sud anche le donne migrano."

42. Cited in Ingram, "The Aesthetics and Ethics of Eco-Film Criticism," 45.

43. MacDonald, "The Ecocinema Experience," 26.

44. The film's full title is *Il Sibilo lungo della taranta: Musiche e poesie sui percorsi del ragno del Salento and Ju tarramutu* (2006).

45. Congedo, "Dai veleni dell'Ilva alla voglia di riscatto."

46. "Taranto, come L'Aquila, è lo specchio dell'Italia: il risultato delle peggiori scelte politiche che si potessero fare. E' una città che paga le conseguenze di un modello di sviluppo industriale completamente in conflitto con l'ambiente naturale e la salute, che tutti hanno sostenuto finchè hanno potuto. . . . Mi ha fatto riflettere il mea culpa dei cittadini e degli stessi operai che hanno ammesso di avere tenuto gli occhi chiusi per troppo tempo" (ibid.).

47. "La situazione sembra disperata ma Taranto non ha solo una bellezza perduta. C'è una bellezza, ancora attuale, che merita di essere valorizzata" (ibid.).

48. "portano avanti una massiccia azione di contro-informazione nei confronti di una popolazione passiva, abituata storicamente al silenzio e all'accettazione" (Perniola, "Poetiche dell'invisibile," 22).

49. I quote from the English subtitles that accompany the film.

50. Also present at the end of the film is Michelle Riondino, a Taranto native known nationally for starring in the titular role in the television show *Il Giovane Montalbano*, which aired in 2012 and 2015.

51. Pisanelli, "Buongiorno Taranto."

52. Somigli, *Negli archivi e per le strade*, iv.

53. Personal communication with Grace Zanotto, June 2, 2020.

54. Certeau, *The Practice of Everyday Life*, 108.

55. Safran, *The Medieval Salento*, 57.

56. Ivakhiv, "An Ecophilosophy of the Moving Image," 98.

57. Lefebvre, *Landscape and Film*.

58. Alaimo, "Trans corporeality," 436.

59. "L'idea di base dalla quale è nato il progetto era che quella Fabbrica di sofferenza e inquinamento andasse bloccata, e la nostra rabbia era così forte da essere pronti a farla esplodere. Infatti . . . a modo nostro lo abbiamo fatto!!" (personal communication with Grace Zanotto, June 2, 2020)

60. Cubitt, "Everybody Knows This Is Nowhere," 279.

61. Bozak, *The Cinematic Footprint*, 5.

62. Past, *Italian Ecocinema*, 57.

63. Di Bianco, "Ecocinema Ars et Praxis, Alice Rohrwacher's Lazzaro Felice," 154.

64. Past, *Italian Ecocinema*, 57.

65. Di Bianco, "Ecocinema Ars et Praxis, Alice Rohrwacher's Lazzaro Felice," 155.

66. "The Apulia Film Commission Foundation, set up under regional legislation, was established in 2007 with the aim of attracting audiovisual production companies to the area through the good value of its facilities, the professionalism of its skilled personnel and the lowering of costs for travel, cast, crew and location scouting in the region" (http://en.apuliafilmcommission.it/about-us/company-profile).

6. READING LANDSCAPE

1. Latour, "On Actor-Network Theory," 7.

2. Herman, "Narratology as a Cognitive Science," 9.

3. James and Morel, "Ecocriticism and Narrative Theory," 363.

4. Spirn, *The Language of* Landscape, 16; Antrop, "A Brief History of Landscape Research," 12; see also Jackson, *Discovering the Vernacular Landscape*.

5. Schama, *Landscape and Memory*, 10.

6. Biggell and Chang, "The Meanings of Landscape," 85, my emphasis.

7. Jackson describes landscape as "a synthetic space, a man-made system of spaces superimposed on the face of the land, functioning and evolving not according to natural laws but to serve a community . . . thus a space deliberately created to speed up or slow down the process of nature" (*Discovering the Vernacular Landscape*, 8). I push back against Jackson's claim that landscapes do not evolve according to natural laws—for the nonhuman surely still has agency in any given landscape—but his description of a deliberate manipulation of the time of the nonhuman is entirely apt in the case of the Oak Forest, in which intensive human-directed bioremediation was undertaken, and park planners chose fast growing species so as to quickly offer a filled in green space for residents, thus creating the "forest" in a way that could never occur naturally in the nonhuman world.

8. Pagano, "Reclaiming Landscape," 402.

9. See Seger, *Landscapes in Between*, for a discussion of interstitial Italian landscapes.

10. Powell, "Walking Refrains for Storied Movement."

11. Spirn, *The Language of Landscape*, 15.

12. Past, *Italian Ecocinema*, 10.

13. An official publication of the Lombardy Region edited by Mario Di Fidio and published in 2000 notes that estimates range from 300 grams to 130 kilograms (24).

14. Conti, "Seveso sei anni dopo" (with thanks to the Circolo Legambiente Laura Conti Seveso for sharing their archives). Conti does mention one "mysterious woman" who knows where the material ended up: an unnamed employee of the firm responsible for transporting the toxic materials.

15. "affermazione del paesaggio come patrimonio da conservare e trasmettere alle generazioni a venire. Un paesaggio in cui il tratto caratteristico era riconosciuto nella presenza di ampie aree boschive" (Centemeri, *Ritorno a Seveso*, 129).

16. Fratter writes that "most of the contaminated material is made up of the top level of soil from the most contaminated area (Zone A), along with parts of houses, machinery, personal objects, dead animals, and some of the initial cleanup materials" (Fratter, *Seveso*, 31).

17. Cohen, *Stone*, 107.

18. See http://www.soilhealth.com/soil-health/biology/degredation.htm#2.

19. See Di Fidio, *Il "Bosco delle Querce" di Seveso e Meda*, 39.

20. Past, *Italian Ecocinema*, 61.

21. "un'area forestale di origine antropica, la cui composizione vegetale è ispirata ai vicini boschi originari, costituiti in prevalenza di Farnia, Pino Silvestre, Betulla, Carpino bianco, Ontano nero e Salice bianco" (Fratter, *Seveso*, 32).

22. "bosco ad evoluzione spontanea, caratterizzato da un ricco sottobosco, che costituisce un'area a vocazione naturalistica ad accesso limitato" (ibid.).

23. Carr, *What Is History?*, 3.

24. "Avevamo pensato a un lavoro in cui dare l'opportunità alle persone di tirare fuori la loro rabbia, il loro dolore. Perché lì a Seveso si sente che c'è qualcosa nel corpo sociale . . . qualcosa che non va . . . Non c'è stato verso: si è visto subito che di certe cose non si doveva parlare" (Fratter, *Seveso*, 40).

25. "la questione più spinosa—i danni alla salute in primis" (Centemeri, *Ritorno a Seveso*, 176).

26. "la memoria collettiva che prende forma nei pannelli del Bosco delle Querce insiste sulla capacità che Seveso ha dimostrato di mettersi il disastro alle spalle, ma ignorando le questioni che hanno diviso e continuano a dividere. A partire di una sorta di visione positiva dell'incidente, essa cerca di fare in modo che il disastro entri nella definizione di un'identità locale che valorizza insieme l'associazionismo, il radicamento nel territorio, e la tutela dell'ambiente" (ibid., 180).

27. "Perché non fu una comunità viva, ma furono tante piccole comunità, quindi la parte cattolica, i primi ambientalisti, Laura Conti, c'è, no, ma non ci fu coesione con le diverse posizioni, anzi, ci fu contrapposizione . . . allora è stata una mediazione necessaria, anche per riportare le persone al parco" (Massiliano Fratter, interview by author, 2015).

28. "Nulla è fuori posto, tutto è 'perfetto'. Chi non conosce le origini dell'area viene positivamente colpito dalla bellezza della zona che in questo modo non rappresenta il 'luogo della memoria' del 'caso Seveso'" (Fratter, *Seveso*, 39).

29. "non c'era traccia di memoria di quello che era sucesso in quel bosco. Abbiamo pensato che stava succedendo qualcosa che non andava bene, cioè che sarebbe aperta, normalizzata una situazione senza la memoria di . . . quello che c'è stato sotto quel bosco" (Circolo Legambiente, interview by author, 2015, my emphasis).

30. Davis, "Rewilding Distilled."

31. Sadeghian and Vardanyan, "A Brief Review," 120.

32. Jackson, *Discovering the Vernacular Landscape*, 13.

33. Hall, Ham, and Lackey, "Comparative Evaluation," 18.

34. I thank Damiano Benvegnù for introducing me to the work of Eugenio Turri.

35. "Ma un modo per appropriarsi, nella lettura del paesaggio, della condizione naturale, di assumerla come componente fondamentale . . . starebbe proprio nel considerare il segno umano come il risultato di un rapporto comunicativo tra uomo e ambiente naturale" (Turri, *Il paesaggio come teatro*, 165).

36. "lo spazio locale, il territorio minimo delle nostre passeggiate, si inserisce in un contesto spaziale e temporale più ampio—del quale anche il nostro camminare tra le colline deve tener conto" (Turri, *Il paesaggio come teatro*, 201).

37. Iovino, *Ecocriticism and Italy*, 26–27.

38. I refer to the twelfth biennial conference of the Association for the Study of Literature and the Environment, "Rust/Resistance: Works of Recovery," held in Detroit, Michigan, in June 2017. I thank Damiano Benvegnù, Enrico Cesaretti, Serenella Iovino, and Elena Past for the productive dialogues that emerged as we crafted this panel together.

39. Andre, Hausman, and Guerriero, "*Cannabis sativa*: The Plant of the Thousand and One Molecules."

40. Raskin et al., "Bioconcentration of Heavy Metals by Plants."

41. United Nations Environment Programme (UNEP) International Environmental Technology Centre, United States Environmental Protection Agency, and Environment Canada, *Phytoremediation: An Environmentally Sound Technology for Pollution Prevention, Control and Remediation: An Introductory Guide for Decision-Makers*, Freshwater Management Series No. 2, 2000.

42. Gutterman, "Back to Chernobyl."

43. Beans, "Core Concept."

44. See Linger, Ostwald, and Haensler, "*Cannabis sativa L.*"

45. "Papà ha sempre amato e tutt'ora ama la terra, gli uliveti, gli animali. . . . ama le Masserie soprattutto di colore bianco . . . ama girare per Aziende . . . ama l'idea dell'allevamento . . . il senso della gestione dei dipendenti" (Fornaro, "Masseria Carmine–Taranto").

46. Information about the Masseria Carmine's current activities, including guided visits and horseback riding lessons, can be found at https://www.masseriacarmine.it/.

47. "l'ennesimo tentativo per resistere e per dimostrare che la nostra splendida terra può continuare a vivere!!" (ibid.).

48. See, for example, the following article in *Vice:* "Intorno all'Ilva di Taranto stanno coltivando la cannabis per bonificare i terreni," https://news.vice.com/it/article/canapa-bonifica-area-terreni-taranto.

49. Noel Gazzano: Visionary Art and Anthropology for Social Change, http://www.noelgazzano.com.

50. Clips and interviews can be viewed at https://vimeo.com/user46489251.

51. For an overview of Puglia's Xylella fastidiosa problem, see Borunda, "Italy's Olive Trees Are Dying," *National Geographic*, August 10, 2018.

52. "Io t'amo, terra mia . . . e anche dire che t'amo sciocco, perchè sono tua. Le tue zolle sono la mia carne, i tuoi tronchi d'ulivo la mia carne, e il veleno con cui t'ammorbano terra mia, io me lo sento nelle vene, e m'arriva fino in bocca, mentre prima c'era solo zucchero e mandorle, e il gusto più amaro che conoscevamo era la cicoria resta."

53. See Ensor, "Spinster Ecology," 409–35.

54. Noel Gazzano: Visionary Art and Anthropology for Social Change, http://www.noelgazzano.com/.

55. Fisher-Lichte, *The Transformative Power of Performance*, 207. Thanks again to Damiano Benvegnù for also steering me toward Fisher-Lichte.

56. Rodrìguez, "Where Can Walking Be Taking Me?," 198.

57. Dumit and O'Connor, "The Senses and Sciences of Fascia," 50.

58. Ammostro collective, interview by author, 2018.

59. Armiero et al., "Toxic Bios," 8.

60. Sommer, *The Work of Art in the World*, 50–51.

61. Ibid.

62. Past, *Italian Ecocinema*, 5.

63. Povinell, *Economies of Abandonment*, 134.

64. Ibid., 7.

FIRST PERSON: THE AMMOSTRO ARTIST COLLECTIVE

1. A *rosta* is a Late Baroque architectural design element made of wrought iron and situated about a building's main doorway.

Bibliography

Adam, Barbara. *Timescapes of Modernity: The Environment and Invisible Hazards*. New York: Routledge, 1998.
Agency for Toxic Substances and Disease Registry (ATSDR). "Public Health Statement Chlorinated Dibenzo-p-Dioxins (CDDs)." December 1998.
———. "Toxicological Profile for Chlorinated Dibenzo-p-dioxins (CDDs)." Atlanta, GA: U.S. Department of Health and Human Services, Public Health Service (1998). https://www.atsdr.cdc.gov/toxprofiles/tp.asp?id=366&tid=63.
Agenzia Nazionale Stampa Associata. "Ex Ilva: Studio Sentieri, 600 bimbi malformati a Taranto." ANSA.it. June 1, 2019. http://www.ansa.it/canale_saluteebenessere/notizie/sanita/2019/06/01/ex-ilva-bonelli-verdi-nascosti-dati-su-600-bimbi-malformati_b966b4e1-30ce-4a1d-8731-b109c4fc6bb7.html.
Alaimo, Stacy. *Bodily Natures: Science, Environment and the Material Self*. Bloomington: Indiana University Press, 2010.
———. "Trans corporeality." In *The Posthuman Glossary*, edited by Rosi Braidotti and Maria Hlavajova, 435–38. London: Bloomsbury, 2018.
Alaimo, Stacy, and Susan Hekman, eds. *Material Feminisms*. Bloomington: Indiana University Press, 2008.
Andre, Christelle M., Jean-Francois Hausman, and Gea Guerriero. "*Cannabis sativa*: The Plant of the Thousand and One Molecules." *Frontiers in Plant Science* 7, no. 19 (2016). https://www.ncbi.nlm.nih.gov/pmc/articles/PMC4740396/.
Angelone, Anita, and Clarissa Clò. "Other Visions: Contemporary Italian Documentary Cinema as Counter-Discourse." *Studies in Documentary Film* 5, no. 2–3 (2011).
ANSA. "Dioxins Found in Taranto Breast Milk." July 3, 2018. http://www.ansa.it/english/news/general_news/2014/05/30/dioxins-found-in-taranto-breast-milk_0b87696f-7147-4948-9118-f1d52b96acfd.html.
———. "Ex Ilva: Studio Sentieri, 600 bimbi malformati a Taranto." May 16, 2021. http://www.ansa.it/canale_saluteebenessere/notizie/sanita/2019/06/01/ex-ilva-bonelli-verdi-nascosti-dati-su-600-bimbi-malformati_b966b4e1-30ce-4a1d-8731-b109c4fc6bb7.html.
Antonello, Pierpaolo. "Dispatches from Hell: *Gomorra* by Matteo Garrone." In *Mafia Movies: A Reader*, edited by Dana Renga. Toronto: Toronto University Press, 2011.

Antrop, Marc. "A Brief History of Landscape Research." In *The Routledge Companion to Landscape Studies*, edited by Peter Howard, Ian Thompson, Emma Waterton, and Mick Atha. London: Routledge, 2018.

Appadurai, Arjun. *Modernity at Large: Cultural Dimensions of Globalization*. Minneapolis: University of Minnesota Press, 1998.

Argentina, Cosimo. *Vicolo dell'acciaio*. Rome: Fandango Libri, 2010.

Armiero, Marco. "An Environmental Historian among Activists: The Political, the Personal, and a Project of Guerrilla Narrative." In *Italy and the Environmental Humanities: Landscapes, Natures, Ecologies*, edited by Serenella Iovino, Enrico Cesaretti, and Elena Past. Charlottesville: University of Virginia Press, 2018.

———. "Introduzione. Raccontare è resistere." In *Teresa e le altre: Storie di donne nella terra dei fuochi*, edited by Marco Armiero, 9–17. Milan: Jaca, 2014.

Armiero, Marco, Thanos Andritsos, Stefania Barca, Rita Brás, Sergio Ruiz Cauyela, Çağdaş, Dedeoğlu, Marica Di Pierri, Lúcia de Oliveira Fernandes, Filippo Gravagno, Laura Greco, Lucie Greyl, Ilenia Iengo, Julia Lindblom, Felipe Milanez, Sérgio Pedro, Giusy Pappalardo, Antonello Petrillo, Maurizio Portaluri, Elisa Privitera, Ayşe Ceren Sari, and Giorgos Velegrakis. "Toxic Bios: Toxic Autobiographies—A Public Environmental Humanities Project." *Environmental Justice* 12, no. 1 (2019): 7–11.

ARPA Puglia. "Cronologia emissioni da impianti agglomerazione ILVA 1994–2011." September 14, 2015. www.arpa.puglia.it.

Attino, Tonio. *Generazione Ilva: Gli ulivi, le industrie, il book, il declino, l'inquinamento. La tragica parabola di una terra illusa dall'acciaio, tradita dallo Stato*. Lecce: Salento Books, 2013.

Auyero, Javier, and Deborah Swistun. "The Social Production of Toxic Uncertainty." *American Sociological Review* 73, no. 3 (2008): 357–79.

Barad, Karen. *Meeting the Universe Halfway: Quantum Physics and the Entanglement of Matter and Meaning*. Durham, NC: Duke University Press, 2007.

Barca, Stefania. "Laboring the Earth: Transnational Reflections on the Environmental History of work." *Environmental History* 19, no. 1 (2014): 3–27.

———. "Lavoro, corpo, ambiente: Laura Conti e le origini dell'ecologia politica in Italia." *Ricerche storiche* 41, no. 3 (2011): 541–50.

———. "Work, Bodies, Militancy: The 'Class Ecology' Debate in 1970s Italy." In *Powerless Science? Science and Politics in a Toxic World*, edited by Soraya Boudia and Nathalie Jas, 115–33. New York: Berghahn, 2014.

Barca, Stefania, and Emanuele Leonardi. "Working-Class Communities and Ecology: Reframing Environmental Justice around the Ilva Steel Plant in Taranto (Apulia, Italy)." In *Class, Inequality and Community Development*, edited by Mae Shaw and Marjorie Mayo, 57–75. Bristol, UK: Bristol University Press, 2016.

———. "Working Class Ecology and Union Politics: A Conceptual Topology." *Globalizations* 15, no. 4 (2018): 487–503.

Beans, Carolyn. "Core Concept: Phytoremediation Advances in the Lab but Lags in the Field." *PNAS* 114, no. 29 (2017): 7475–77.

Ben-Ghiat, Ruth. *Italian Fascism's Empire Cinema*. Bloomington: Indiana University Press, 2015.
Bennett, Jane. *Vibrant Matter: A Political Ecology of Things*. Durham, NC: Duke University Press, 2009.
Beck, Eckardt C. "The Love Canal Tragedy." *EPA Journal* (US Environmental Protection Agency). January 1979. https://archive.epa.gov/epa/aboutepa/love-canal-tragedy.html.
Berger, Bruce. "Telling It Slant: The Value of Narrative Indirection." In *What's Nature Worth? Narrative Expressions of Environmental Values*, edited by Terre Satterfield and Scott Slovic, 234–42. Salt Lake City: University of Utah Press, 2004.
Berger, John. *About Looking*. New York: Vintage, 1980.
Bertazzi, Pier A., Ilaria Bernucci, Gabriella Brambilla, Dario Consonni, and Angela C. Pesatori. "The Seveso Studies on Early and Long-Term Effects of Dioxin Exposure: A Review." *Environmental Health Perspectives* 106 (Suppl 2) (1998): 625–33. Web.
Bertozzi, Marco. *Storia del documentario italiano*. Milan: Marsilio, 2008.
Biacchessi, Daniele. *La fabbrica dei profumi: La verità su Seveso, l'Icmesa, la diossina*. Milan: Baldini Castoldi Dalai, 1995.
Bianco, Giuliana, Rosalia Zianni, Giuseppe Anzillotta, Achille Palma, Vincenzo Vitacco, Laura Scrano, and Tomasso R. I. Cataldi. "Dibenzo-*p*-Dioxins and Dibenzofurans in Human Breast Milk Collected in the Area of Taranto (Southern Italy): First Case Study." *Analytical and Bioanalytical Chemistry* 405 (2013): 2405–10. https://doi.org/10.1007/s00216-013-6706-7.
Biehl, Joao, and Amy Moran-Thomas. "Symptoms: Subjectivities, Social Ills, Technologies." *Annual Review of Anthropology* 38, no. 1 (2009): 267–88.
Biggell, Werner, and Cheng Chang. "The Meanings of Landscape: Historical Development, Cultural Frames, Linguistic Variation, and Antonyms." *Ecozon@* 5, no. 1 (2014): 84–103.
Blanchot, Maurice. *The Writing of the Disaster*. Translated by Ann Smock. Lincoln: University of Nebraska Press, 1995.
Boldrini, Sante. "Intervista a Laura Conti." *Sabato Sera*. Accessed May 29, 2015, at Laura Conti Archives, Fondazione Biblioteca Archivio Luigi Micheletti.
Bondavalli, Simona. "Broadcasting Human-Animal Friendship: RAI Television's *I nostri amici* and Early Italian Ecodocumentaries." *The Italianist* 40, no. 2 (2020): 190–202.
Borunda, Alejandra. "Italy's Olive Trees Are Dying: Can They Be Saved?" *National Geographic*, August 10, 2018.
Bozak, Nadia. *The Cinematic Footprint: Lights, Camera, Natural Resources*. New Brunswick, NJ: Rutgers University Press, 2011.
Braidotti, Rosi. *The Posthuman*. Cambridge: Polity, 2013.
———. "A Theoretical Framework for the Critical Posthumanities." *Theory, Culture and Society* 36, no. 6 (2019): 31–61.
Bruzzi, Stella. *New Documentary*. 2nd ed. New York: Routledge, 2006.

Buell, Lawrence. "Toxic Discourse." *Critical Inquiry* 24, no. 3 (1998): 639–65.
Calcagno, Paolo. "Due ragazzi 'intossicati' dalla diossina menzognera." *Corriere della Sera* September 1988 (1 page). Accessed May 29, 2015, at Laura Conti Archives, Fondazione Biblioteca Archivio Luigi Micheletti.
Caminati, Luca, and Mauro Sassi, "Notes on the History of Italian Nonfiction Film." In *A Companion to Italian Cinema*, edited by Frank Burke, 361–73. Hoboken, NJ: Wiley, 2017.
Caprino, Carlo. "*Adesso tienimi*, il libro d'esordio di Flavia Piccinni riletto dieci anni dopo." *Grottaglie in rete*, August 6, 2017. https://www.grottaglieinrete.it/it/adesso-tienimi-romanzo-desordio-flavia-piccinni-riletto-dieci-anni/.
Caracciolo, Marco. *Strange Narrators in Contemporary Fiction: Explorations in Readers' Engagement with Characters*. Lincoln: University of Nebraska Press, 2016.
Carr, E. H. *What Is History?* London: Harmondsworth, 1964.
Castelli, Valeria. "The Filmmaker Is Present: Performance, Ethos, and Politics in *In viaggio con Cecilia* and *Io sto con la sposa*." *The Italianist* 38, no. 2 (2018): 235–57.
Cederna, Antonio. "Taranto in balia dell'Italsider." *Corriere della Sera*, April 13, 1972, 3.
Celati, Gianni. *Verso la foce*. 3rd ed. Milan: Feltrinelli, 2002.
Centemeri, Laura. *Ritorno a Seveso: Il danno ambientale, il suo riconoscimento, la sua riparazione*. Milan: Mondadori, 2006.
———. "The Seveso Disaster Legacy." In *Nature and History in Modern Italy*, edited by Marco Armiero and Marcus Hall, 251–73. Athens: Ohio University Press, 2010.
Cesaretti, Enrico. *Elemental Narratives: Reading Environmental Entanglements in Modern Italy*. University Park: Pennsylvania State University Press, 2020.
Chapman, Jane. *Documentary in Practice: Filmmakers and Production Choices*. Hoboken, NJ: Wiley, 2007.
Chatman, Seymour. "What Can We Learn from Contextualist Narratology?" *Poetics Today* 11, no. 2 (1990): 309–28.
Chen, Mel. *Animacies: Biopolitics, Racial Mattering, and Queer Affect*. Durham, NC: Duke University Press, 2012.
Cohen, Jeffrey Jerome. *Stone: An Ecology of the Inhuman*. Minneapolis: University of Minnesota Press, 2015.
Colella, Gaetano. *Capatosta*. 2014. Script shared by the author.
Congedo, Alessandra. "Dai veleni dell'Ilva alla voglia di riscatto Intervista a Paolo Pisanelli, regista del film 'Storie di Taranto.'" *Inchiostro Verde*, November 16, 2012. https://www.inchiostroverde.it/2012/11/16/dai-veleni-dellilva-alla-voglia-di-riscatto-intervista-a-paolo-pisanelli-regista-del-film-storie-di-taranto/.
Consonni, Dario, Angela C. Pesatori, Carlo Zocchetti, Raffaela Sindaco, Luca Cavalieri D'Oro, Maurizia Rubagotti, and Pier Alberto Bertazzi. "Mortality in a Population Exposed to Dioxin after the Seveso, Italy, Accident in 1976: 25 Years of Follow-up." *American Journal of eEpidemiology* 167, no. 7 (2008): 847–58.
Conti, Laura. *Cecilia e le streghe*. Turin: Einaudi, 1963.
———. *Che cos'è l'ecologia: capitale, lavoro e ambiente*. Milan: G. Mazzotta, 1965.
———. *La condizione sperimentale*. Milan: Mondadori, 1965.
———. *Il dominio sulla materia*. Milan: Mondadori, 1973.

———. *Sesso ed educazione.* Rome: Riuniti, 1970.

———. "Seveso sei anni dopo: Solo una misteriosa signora sa dov'è finita la diossina." *L'Unità,* October 17, 1982, 22. https://archivio.unita.news/assets/main/1982/10/17/page_024.pdf.

———. *Una lepre con la faccia di bambina.* Milan: Riuniti, 1978.

———. *Visto da Seveso.* Milan: Riuniti, 1977.

Cubitt, Sean. "Everybody Knows This Is Nowhere: Data Visualization and Ecocriticism." In *Ecocinema Theory and Practice,* edited by Stephen Rust, Salma Monani, and Cubitt, 279–96. New York: Routledge, 2012.

Davis, John. "Rewilding Distilled." *Rewilding Earth,* July 21, 2018. https://rewilding.org/rewilding-distilled/.

Certeau, Michel de. *The Practice of Everyday Life.* Translated by Steven Rendall. Berkeley: University of California Press, 1984.

Desiati, Mario. "Cosimo Argentina, Maschio adulto solitario." April 27, 2008. www.mannieditori.it.

Di Bianco, Laura. "Ecocinema Ars et Praxis, Alice Rohrwacher's Lazzaro Felice." *The Italianist* 40, no. 2 (2020): 151–64.

Donly, Corinne. "Toward the Eco-Narrative: Rethinking the Role of Conflict in Storytelling." *Humanities* 6, no. 2: 17 (2017). https://doi.org/10.3390/h6020017.

Donnarumma, Raffaele, *Ipermodernità: Dove va la narrativa contemporanea.* Bologna: Il Mulino, 2014.

D'Ovidio, Marianna. "A Taranto di muri ce ne sono tanti." *cheFare,* June 29, 2016. https://www.che-fare.com/taranto-muri/.

Dumit, Joseph. *Drugs for Life: How Pharmaceutical Companies Define Our Health.* Durham, NC: Duke University Press, 2012.

Dumit, Joseph, and Kevin O'Connor. "The Senses and Sciences of Fascia." In *Sentient Performativities of Embodiment: Thinking alongside the Human,* edited by Lynette Hunter, Elisabeth Krimmer, and Peter Lichtenfels, 35–54. Lanham, MD: Lexington, 2016.

Ensor, Sarah. "Spinster Ecology: Rachel Carson, Sarah Orne Jewett, and Nonreproductive Futurity." *American Literature* 84, no. 2 (2012): 409–35.

Environmental Protection Agency. "Gulf Oil Spill: Assessment of Dioxin Emissions from *in situ* Oil Burns." May 16, 2021. Washington, DC: Environmental Protection Agency.

Environmental Working Group, "Dioxin Timeline." July 13, 2010. https://www.ewg.org/release/dioxin/home.

Ferraris, Maurizio. "Il Ritorno al pensiero forte." *la Reppublica,* August 8, 2011.

Fetters, Ashley. "The Tampon: A History. The Cultural, Political, and Technological Roots of a Fraught Piece of Cotton." *The Atlantic,* June 1, 2015.

Fisher-Lichte, Erika. *The Transformative Power of Performance: A New Aesthetics.* Translated by Saskya Iris Jain. London: Routledge, 2008.

Fondazione Cineteca di Bologna. "Cecilia Mangini e Lino del Fra." N.d. http://www.cinetecadibologna.it/biblioteca/patrimonioarchivistico/fondo_mangini_delfra.

Fornaro, Rosanna. "Blog." *Masseria Carmine-Taranto.* https://www.masseriacarmine.it/blog-1/.

Foschini, Giuliano. "Pecore tosiche abbattute: 'Malati anche noi pastori.'" *la Repubblica, Bari,* December 16, 2008. https://bari.repubblica.it/dettaglio/pecore-tossiche-abbattute:-malati-anche-noi-pastori/1562567.

———. *Quindici passi.* Rome: Fandango Libri, 2009.

Fratter, Massimiliano. *Seveso: Memories da sotto il bosco.* Milan: Legambiente, 2006.

Garambois, Silvia. "Seveso fa ancora paura." *l'Unità,* September 19, 1988.

Gazzano, Noel. *Noel Gazzano: Visionary Art and Anthropology for Social Change.* https://www.noelgazzano.com/.

Genette, Gérard. *Narrative Discourse: An Essay in Method.* Translated by Jane E. Lewin. Ithaca, NY: Cornell University Press, 1980.

Giannì, Roberto, and Anna Migliaccio. "Taranto, oltre la crisi." *Meridiana: Rivista di storia e scienze sociali:* 85, no. 1 (2016): 159.

Gutterman, Lila. "Back to Chernobyl." *New Scientist,* no. 2181, April 10, 1999. https://www.newscientist.com/article/mg16221810-900-back-to-chernobyl/.

Hall, Troy E., Sam H. Ham, and Brenda K. Lackey. "Comparative Evaluation of the Attention Capture and Holding Power of Novel Signs Aimed at Park Visitors." *Journal of Interpretation Research* 15, no. 1 (2015): 15–36.

Hayward, Eva. "More Lessons from a Starfish: Prefixial Flesh and Transspeciated Selves." *Women's Studies Quarterly* 36, no. 3/4 (Fall–Winter 2008): 64–85.

Heise, Ursula K. "Eco-narrative." In *Routledge Encyclopedia of Narrative Theory,* edited by David Herman, Manfred Jahn, and Marie-Laure Ryan, 129–30. London: Routledge, 2005.

Herman, David. "Narratology as a Cognitive Science." *Image & Narrative* 1, no. 1 (2000). http://www.imageandnarrative.be/inarchive/narratology/davidherman.htm.

———. *Story Logic: Problems and Possibilities of Narrative.* Lincoln: University of Nebraska Press, 2002.

Houppert, Karen. *The Curse: Confronting the Last Unmentionable Taboo: Menstruation.* New York: Farrarr, Straus and Giroux, 1999.

Hunter, Lynette, Elisabeth Krimmer, and Peter Lichtenfels, eds. *Sentient Performativities of Embodiment: Thinking alongside the Human.* Lanham, MD: Lexington, 2016.

Hunter, Russ. "Nightmare Cities: Italian Zombie Cinema and Environmental Discourses." In *Screening the Undead,* edited by Leon Hunt, Sharon Lockyer, and Milly Williamson, 112–30. London: I. B. Taurus, 2013.

Ingram, David. "The Aesthetics and Ethics of Eco-Film Criticism." In *Ecocinema Theory and Practice,* edited by Stephen Rust, Salma Monani, and Cubitt, 43–62. New York: Routledge, 2012.

Iovino, Serenella. *Ecocriticism and Italy.* London: Bloomsbury, 2016.

———. "Toxic Epiphanies: Dioxin, Power, and Gendered Bodies in Laura Conti's Narratives on Seveso." *International Perspectives in Feminist Ecocriticism,* 37–55, edited by Greta Gaard, Simon Estok, and Serpil Oppermann. New York: Routledge, 2013.

Iovino, Serenella, and Serpil Oppermann, eds. *Material Ecocriticism*. Bloomington: Indiana University Press, 2104.
Ivakhiv, Adrian. "An Ecophilosophy of the Moving Image." In *Ecocinema Theory and Practice*, edited by Stephen Rust, Salma Monani, and Sean Cubitt, 87–106. New York: Routledge, 2012.
Jackson, J. B. *Discovering the Vernacular Landscape*. New Haven, CT: Yale University Press, 1984.
James, Erin. *The Storyworld Accord: Econarratology and Postcolonial Narratives*. Lincoln: University of Nebraska Press, 2015.
James, Erin, and Eric Morel. "Ecocriticism and Narrative Theory: An Introduction." *English Studies* 99, no. 4 (2018): 355–65.
Kington, Tom. "Italian Town Fighting for Its Life over Polluting Ilva Steelworks." *The Guardian*, August 17, 2012. https://www.theguardian.com/world/2012/aug/17/italy-ilva-steelworks-cancer-pollution.
Knickerbocker, Scott. *Ecopoetics: The Language of Nature, The Nature of Language*. Amherst: University of Massachusetts Press, 2012.
Latour, Bruno. "On Actor-Network Theory: A Few Clarifications Plus More Than a Few Complications." *Soziale Welt* 47 (1996): 369–81.
Lefebvre, Martin, ed. *Landscape and Film*. New York: Routledge, 2006.
Lehtimäki, Markku. "Natural Environments in Narrative Contexts: Cross-Pollinating Ecocriticism and Narrative Theory." *StoryWorlds: A Journal of Narrative Studies* 5 (2013): 119–41.
Liboiron, Max, Manuel Tironi, and Nerea Calvillo, "Toxic Politics: Acting in a Permanently Polluted World." *Social Studies of Science* 48 (2018): 331–49.
Linger, P., A. Ostwald, and J. Haensler, "*Cannabis sativa* L. Growing on Heavy Metal Contaminated Soil: Growth, Cadmium Uptake and Photosynthesis." *Biologia Plantarum* 49 (2005): 567–76.
MacDonald, Scott. "The Ecocinema Experience." In *Ecocinema Theory and Practice*, edited by Stephen Rust, Salma Monani, and Sean Cubitt, 17–42. New York: Routledge, 2012.
Mangini, Cecilia. "Film Notes—Essere Donne." Il Cinema Ritrovato. N.d. https://festival.ilcinemaritrovato.it/en/film/essere-donne/.
Mascherpa, Barbara. *La stampa quotidiana e la catastrophe di Seveso: Verità e falsità dei giornali di fronte al problema 'aborto.'* Milan: Vita e Pensiero, 1990.
Mastrodonato, Luigi. "Intorno all'Ilva di Taranto stanno coltivando la cannabis per bonificare i terreni." *Vice*, February 4, 2016.
Missero, Dalilla. "Cecilia Mangini: A Counterhegemonic Experience of Cinema." *Feminist Media Histories* 2, no. 3 (2016): 54–72.
Mosca, Raffaele Palumbo. "New Realisms or Return to Ethics?" In *Encounters with the Real in Contemporary Italian Literature and Cinema*, 47–68, edited by Loredana di Martino and Pasquale Verdicchio. Newcastle upon Tyne: Cambridge Scholars Publishing, 2017.
Nash, Kate. "Documentary-for-the-Other: Relationships, Ethics and (Observational) Documentary." *Journal of Mass Media Ethics*, 26, no. 3 (2011): 224–39.

Needham, L. L., et al. "Serum Dioxin Levels in Seveso, Italy, Population in 1976." *Teratogenesis, Carcinogenesis, and Mutagenesis* 17, no. 4–5 (1997): 225–40.

Nichols, Bill. *Introduction to Documentary.* Bloomington: Indiana University Press, 2011.

Nixon, Rob. *Slow Violence and the Environmentalism of the Poor.* Cambridge, MA: Harvard University Press, 2011.

Nussbaum, Martha. *Cultivating Humanity: A Classical Defense of Reform in Liberal Education.* Cambridge, MA: Harvard University Press, 1997.

O'Leary, Alan. "What Is Italian Cinema?" *California Italian Studies* 7, no. 1 (2017): 1–26.

Pagano, Tullio. "Reclaiming Landscape." *Annali d'Italianistica* 29 (2011): 401–16.

Pasolini, Pierpaolo. *Il viaggio jonico: Da Taranto a Leuca.* Lecce: Kurumuny Edizioni, 2017.

Past, Elena. *Italian Ecocinema beyond the Human.* Bloomington: Indiana University Press, 2019.

———. "'Trash Is Gold': Documenting the Ecomafia and Campania's Waste Crisis." *ISLE: Interdisciplinary Studies in Literature and the Environment* 20, no. 3 (2013): 597–621.

Peacelink. https://www.peacelink.it/ecologia/a/40532.html.

Perniola, Ivelise. "Poetiche dell'invisibile: Il Rimosso nel Cinema Documentario Italiano Contemporaneo." *Schermi: Storie e Culture del Cinema e dei Media in Italia* 2, no. 4 (2018): 15–27.

Phelan, James. *Narrative as Rhetoric: Technique, Audiences, Ethics, Ideology.* Columbus: Ohio State University Press, 199.

Piccinni, Flavia. *Adesso tienimi.* Rome: Fazi editore, 2007.

Piccinno, Cristina. "Viaggio in Puglia per Cecilia Mangini e Mariangela Barbanente." *Il Manifesto*, January 26, 2014. https://ilmanifesto.it/viaggio-in-puglia-per-cecilia-mangini-e-mariangela-barbanente/.

Pisanelli, Paolo. "Buongiorno Taranto!" March 11, 2015. http://www.conmagazine.it/2015/03/11/buongiorno-taranto/.

Pollan, Michael. "Some of My Best Friends Are Germs." *New York Times Magazine*, May 15, 2013.

Povinelli, Elizabeth. *Economies of Abandonment: Social Belonging and Endurance in Late Liberalism.* Durham, NC: Duke University Press, 2011.

Powell, Kimberly. "Walking Refrains for Storied Movement." *Cultural Studies—Critical Methodologies* 20, no. 1 (2020): 35–42.

Prudenzano, Angelo. "Arriva da Taranto e dal Salento la 'meglio gioventù' della nuova letteratura italiana." *affaritalini.it*, September 26, 2009. http://www.affaritaliani.it/culturaspettacoli/la_nuova_narrativa_dal_sud_puglia200909.html?refresh_ce.

Raskin, Ilya, P. B. A Nanda Kumar, Slavik Dushenkov, and David E. Salt. "Bioconcentration of Heavy Metals by Plants." *Current Opinion in Biotechnology* 5, no. 3 (1994): 285–90.

Rocca, Francesco. *I giorni della diossina.* supplement to *Quaderni Bianchi* 2 (1980) Milan: Centro Studi A. Grandi, 1980.

Rodrìguez, Àlvaro Ivàn Hernàndez. "Where Can Walking Be Taking Me?" In *Sentient Performativities of Embodiment: Thinking alongside the Human,* edited by Lynette Hunter, Elisabeth Krimmer, and Peter Lichtenfels, 195–204. Lanham, MD: Lexingon, 2016.

Rust, Stephen, Salma Monani, and Sean Cubitt, eds. *Ecocinema Theory and Practice.* New York: Routledge, 2012.

Safran, Linda. *The Medieval Salento: Art and Identity in Southern Italy.* Philadelphia: University of Pennsylvania Press, 2014.

Sadeghian, Mohammad Mehdi, and Zhirayr Vardanyan. "A Brief Review on Urban Park History, Classification and Function." *International Journal of Scientific & Technology Research* 4, no. 11 (2015): 120–24.

Schama, Simon. *Landscape and Memory.* New York: Vintage, 1996.

Scholes, Robert. "Language, Narrative, and Anti-Narrative." In *On Narrative,* edited by W. J. T. Mitchell. Chicago: University of Chicago Press, 1981.

Schoonover, Karl. "Documentaries without Documents? Ecocinema and the Toxic." *NECSUS: European Journal of Media Studies* 2, no. 2 (2013): 483–507.

Seger, Monica. "Aesthetic Engagement: Anna Zanoli's Cultural Programmes for the RAI (1968–1974)." *Italian Studies* 71, no. 4 (2016) 535–49.

———. *Landscapes in Between: Environmental Change in Modern Italian Literature and Film.* Toronto: University of Toronto Press, 2015.

Servizi Parlamentari. "Camera dei Deputati—7–00514—Risoluzione presentata dall'On. Segoni (M5S) ed altri il 6 novembre 2014," 1–2. Accessed September 10, 2015. http://www.serviziparlamentari.com/index.php?option=com_mtree&task=viewlink&link_id=5100&Itemid=2.

Simonetti, Gianluigi. "I nuovi assetti della narrativa italiana (1996–2006)." *Allegoria* 57 (2008): 95–136.

Soles, Carter. "Sympathy for the Devil: The Cannibalistic Hillbilly in 1970s Rural Slasher Films." In *Ecocinema Theory and Practice,* edited by Stephen Rust, Salma Monani, and Sean Cubitt, 233–50. New York: Routledge, 2012.

Somigli, Luca, ed. *Negli archivi e per le strade.* Rome: Aracne, 2013.

Sommer, Doris. *The Work of Art in the World: Civic Agency and Public Humanities.* Durham, NC: Duke University Press, 2014.

Spirn, Anne Whiston. *The Language of Landscape.* New Haven, CT: Yale University Press, 1998.

Thomas, Valerie M., and Thomas G. Spiro. "An Estimation of Dioxin Emissions in the United States." *Toxicological and Environmental Chemistry* 50 (1995): 1–37.

Tuana, Nancy. "Viscous Porosity: Witnessing Katrina." In *Material Feminisms,* edited by Stacy Alaimo and Susan Heckman. Bloomington: Indiana University Press, 2008.

Turri, Eugenio. *Il paesaggio come teatro: Dal territorio vissuto al territorio rappresentato.* Milan: Marsilio, 2006.

United Nations Environment Programme (UNEP) International Environmental Technology Centre, United States Environmental Protection Agency, and Environment Canada. *Phytoremediation: An Environmentally Sound Technology for*

Pollution Prevention, Control and Remediation: An Introductory Guide for Decision-Makers. Freshwater Management Series No. 2 (2002).

U.S. Government. *The Health Risks of Dioxin. Hearing Before the Human Resources and Intergovernmental Relations Subcommittee of the Committee on Government Operations*. House of Representatives, 102nd Cong., June 10, 1992. Washington: U.S. Government Printing Office, 1993.

Vagliasindi, Grazia Maria, and Christiane Gerstetter. *The ILVA Industrial Site in Taranto. In-depth Analysis*, IP/A/ENVI/2015-13, PE 563.471. European Union, 2015.

Vattimo, Gianni, "Dialectics, Difference, Weak Thought." In *Weak Thought*, edited by Vattimo and Pier Aldo Rovatti and translated by Peter Caravetta, 39–52. Albany: State University of New York Press, 2012.

Verdicchio, Pasquale, ed. *Ecocritical Approaches to Italian Culture and Literature: The Denatured Wild*. Lanham, MD: Lexington, 2016.

Willoquet-Maricondi, Paula. *Framing the World: Explorations in Ecocriticism and Film*. Charlottesville: University of Virginia Press, 2010.

World Health Organization. "Dioxins and Their Effects on Human Health." Fact Sheet No. 225/ June 2014. http://www.who.int/mediacentre/factsheets/fs225/en/.

Yoshida, Hideto, Kazuaki Takahashi, Nobuo Takeda, and Shin-ichi Sakai. "Japan's Waste Management Policies for Dioxins and Polychlorinated Biphenyls." *Journal of Material Cycles and Waste Management* 11 (2009): 229–43. https://doi.org/10.1007/s10163-008-0235-z.

Zagaria, Cristina, *Veleno: La battaglia di una giovane donna nella città ostaggio dell'Ilva*. Milan: Sperling and Kupfer, 2013.

FILMOGRAPHY

Belli di Papà. Directed by Guido Chiesa. Medusa, 2015.
Buongiorno Taranto. Directed by Paolo Pisanelli. Big Sur, 2014.
Case sparse: Visioni di case che crollano. Directed by Gianni Celati. Pierrot e la rosa–Stefilm, 2003.
Essere donne. Directed by Cecilia Mangini. Commisione ministrale, 1965.
Il Grande Spirito. Directed by Sergio Rubini. Fandango and Rai Cinema, 2019.
In viaggio con Cecilia. Directed by Cecilia Mangini and Mariangela Barbanente. GA&A Productions, 2013.
Non perdono. Directed by Grace Zanotto and Roberto Marsella. Independent, 2016.
Six Underground. Directed by Michael Bay. Bay Films, Skydance, and Netflix, 2019.
Una lepre con la faccia di bambina. Directed by Gianni Serra. RAI, 1988.

Index

Page numbers in italics refer to illustrations.

Abbott, Lyn, 149
abortion, 2, 28–29, 53, 165; debate over, 29, 44, 47
Abram, David, 145
Acna Montecatini, 34
Actor-network theory, 142
Adam, Barbara, 4, 42–43; *Timescapes of Modernity: The Environment and Invisible Hazards*, 43
Adriatic Sea, 33
Advisory Commission for Health and Ecology (Milan), 37
affect, 1, 31, 46
Agency for Toxic Substances and Disease Registry (ATSDR), 7–9
Agent Orange, 7, 21, 47–48. *See also* dioxin
agriculture, 118–19
Alaimo, Stacy, 10–11, 55, 71, 100, 130
Alexievich, Svetlana, 13
Allegoria (journal), 80
Amati, Daniele, 126–27, 130–31
Ambiente Svenduto (Sold-Out Environment), 67, 163, 174
American Gut Project, 12
Ammostro arts collective, 95, 161, 168–72, 174–77
Angelone, Anita, 109
animacy, 92–94
animals: disposal of, 70; slaughter of, 28, 74–76, 97

Anthropocene, 4
anthropocentrism, 52
Antonello, Pierpaolo, 80
Antonioni, Michelangelo: *Red Desert* (film), 186n18
Antrop, Marc, 143
Appadurai, Arjun, 14
Apulia Film Commission, 134
ArcelorMittal, 2, 67, 96, 173
Argentina, Cosimo, 127; *Il cadetto*, 89; *Cuore di Cuoio*, 89; *Maschio adulto solitario*, 89; *Nud'e cruda: Taranto mon amour*, 89; *Per sempre carnivori*, 89; *L'umano sistema fognario*, 78–79, 89–90; *Vicolo dell'acciaio*, 84, 89–94, 165, 169
Armiero, Marco, 5, 13, 95–96
ARPA Puglia, 67, 84, 97
Asl (Azienda sanitaria locale), 163
Attino, Tonio, 65
Auyero, Javier, 52
Avallone, Silvia: *Acciaio* (Swimming to Elba), 85

Bacchessi, Daniele: *La fabbrica dei profumi: La verità su Seveso, l'Icmesa, la diossina*, 23
Baesler, James, 158
Baldacconi, Rossella, 175
Barad, Karen, 41, 72
Barbanente, Mariangela, 113–17, 119–21; *Ferrhotel* (film), 113; *Sole* (film), 113

Barca, Stefania, 37–38, 41, 68, 185n9
Bari, 90, 96, 112, 119
Barron, Patrick, 44; *Italian Environmental Literature: An Anthology,* 44
Bay, Michael: *Six Underground* (film), 132–35
Beans, Carolyn, 162
Beaumont Bonelli, Filippo di, 163
Belgium, 8
Ben Ghiat, Ruth, 110
Bennett, Jane, 11, 41, 55
Beretta, Gemma, 153
Bertazzi, Pier Alberto, 22
Bertolucci, Bernardo: *La Via del petrolio* (film), 110
Bertozzi, Marco, 110
Biehl, João, 27
Biggell, Werner, 143–44
bioaccumulation, 9–10, 145, 149
bioremediation, 19, 169, 192n7
Blanchot, Maurice, 55
Bocconi University, 169
Bolzano, 37
Bondavalli, Simona, 109
Bozak, Nadia, 111, 133
Braidotti, Rosi, 71
Brereton, Pat, 189n4
Brescia, 24
Brindisi, 114–18
Brockovich, Erin, 82
bubonic plague, 39
Buell, Lawrence, 5
Burgoon, Judee K., 158

Caminati, Luca, 108–9
Canada, 136
Canapuglia, 164
cancer, 9, 28; bladder, 34; breast, 9; death from, 75, 83; elevated rates of mortality attributed to a variety of, 10; lung, 28; pancreatic, 98. *See also* carcinogens
Cannata, Angelo, 82, 101–4, 123
capitalism: extractive, 107, 172; power structure of, 38. *See also* turbocapitalist economy

Caprino, Carlo, 89
Caracciolo, Marco, 16, 117
Carbone, Stefano, 151
carcinogens, 9. *See also* cancer
Carr, E. H., 151, 159–60
Carson, Rachel, 38, 40
Casa Aperta, 32
Cassano, Franco, 172
Castelli, Valeria, 116–17, 190n32
Catholic Church, 29. *See also* Christianity; pope
Cavallaro, Aldo, 24
Cederna, Antonio, 65–66
Celati, Gianni: *Case sparse: Visioni di case che crollano,* 141–42, 160
Cementir, 65
Centemeri, Laura, 18, 61, 148, 151–53; *Ritorno a Seveso,* 23, 25, 27, 142, 180n5
Certeau, Michel de, 84, 127, 143
Cesano Maderno, 22
Cesaretti, Enrico: *Elemental Narratives: Reading Environmental Entanglements in Modern Italy,* 16–17, 85
Chang, Cheng, 143–44
Chapman, Jane, 189n5
chemical warfare, 61
Chen, Mel, 11, 92, 149, 188n46
Chernobyl, 57, 162
Chiesa, Guido: *Belli di Papà,* 134–35
children, 104
chloracne, 8–9, 28, 47–48, 105–6, 181n16
chlorinated dibenzofurans (CDFs), 8. *See also* dioxin; furans
chlorinated dibenzo-p-dioxins (CDDs), 7–8. *See also* dioxin
Christianity, 64. *See also* Catholic Church
Cimmino, Aldo, 147
Cinema del reale (documentary film festival), 112
Circolo Legambiente Laura Conti Seveso, 32–36, 62, 183n45
Cittadini e Lavoratori Liberi e Pensanti, 115, 120
Città Vecchia, 81, 93, 101, 121–29, 133–34, 171, 175. *See also* Taranto
Clò, Clarissa, 109

Cohen, Jeffrey, 13, 148
Cold War: concerns about nuclear contamination and the end of civil society in the, 61; conspiracy theories of the, 61
Colella, Gaetano: *Capatosta* (play), 96
Collodi, Carlo: *Pinocchio*, 43
colonization, 172
Comitato Cittadini e Lavoratori Liberi e Pensanti (Committee of Free and Thinking Citizens and Workers), 69
Confratelli di Carmine (Brothers of the Church of Carmine), 128
CON magazine.it, 124
Conti, Laura, 37–61, 77, 79, 145–46, 152, 193n14; *Ambiente terra*, 37; *Cecilia e le streghe*, 37; *Che cos'è l'ecologia: Capitale, lavoro e ambiente*, 37; *La condizione sperimentale*, 37, 40; *Il dominio sulla materia*, 37; *Una lepre con la faccia di bambina*, 17–18, 31, 38–40, 43–47, 52–56, 60–61, 78; *Questo pianeta*, 37; *Sesso e educazione*, 37; *Visto da Seveso*, 23, 27, 38–44, 52, 61
Corriere della Sera, 24, 181n16
Corriere di Seveso, Il, 39
counter-narrative, 5, 96; plurivocal, 96
Cronauer, Adrian, 121–22
Cubitt, Sean, 132

D'Amato, Bianca Maria, 141
Deleuze, Gilles, 55
De Robertis, Francesco: *Uomini sul fondo (S.O.S. Submarine)*, 109
Desiati, Mario, 89
Desio, 22
Detroit, 171
Dezaki, Satoshi: *Inochi no Chikyuu: Dioxin no natsu* (animation film), 61
diabetes mellitus, 28
Di Bianco, Laura, 133
Dickinson, Emily, 48
Di Maio, Luigi, 67
Di Monopoli, Omar: *Nella perfida terra di Dio* (novel), 79
dioxin, 1–18, 30–34, 42–58, 70–76, 115, 130, 142–45, 149–50, 162, 168; animals with toxicity of, 70, 76, 97, 99, 163; "deviant" nature of, 55; exposure to, 11, 22, 27–28, 31, 39, 48–50, 70, 94, 99, 149; hemp as the opposite of, 167; knowledge about, 60; levels of, 67; ontological complexities posed by, 29; polychlorinated, 7; production of, 9; research with animals on, 9; teratogenic, 29; unknown nature of, 39. *See also* Agent Orange; chlorinated dibenzofurans (CDFs); chlorinated dibenzo-p-dioxins (CDDs); polychlorinated biphenyls (PCBs); toxic chemical compounds
disease: autoimmune, 74; chronic, 79; digestive, 10; perinatal, 10; respiratory, 10; toxins as the cause of, 78, 90–91, 94, 99. *See also* health
dolphins, 63–64
domesticity, 166; daily, 1; sphere of, 88
Donly, Corrine, 15–16, 54, 111
Donnarumma, Raffaele, 81
Dorno, Anna Dora, 77
D'Ovidio, Marianna, 185n1
Dumit, Joseph, 13, 168

ecocinema, 58, 110–11; definition of, 111; Italian, 145; traditional bounds of, 132. *See also* film
eco-corporeal crisis, 1, 6, 18, 78; awareness of, 89. *See also* health crisis; Love Canal crisis; slow violence
ecocritical film studies, 110, 133. *See also* film
eco-dialectic, 167
ecodocumentary, 58. *See also* film
ecology: awareness in film of, 134; catastrophe of, 158; "human-centered," 38, 41
eco-Marxism, 37
EcoMuvi, 134
econarratology, 14–15, 17, 45, 54, 60–61; questions of, 111; studies in, 142
Economist, 39
ecopoetics, 111
eco-terrorism, 131

Ecozon@ (journal), 16
Editori Riuniti, 56
Elba, 66
embodiment: awareness and, 168; of dioxin, 55; discourse of, 87; in the nonhuman world, 38, 168, 172; toxic, 12, 77, 88–89. *See also* landscape; transcorporeality
emissions. *See* industrial emissions
energy: clean, 120; of materials, 17; matter and, 42
Eni, 110; oil refinery of, 65
Ensor, Sarah, 166
environmental humanities, 16, 161
environmental injustice, 5, 96
environmentalism, 149, 165–66
environmental justice, 13; terminology of, 68
environmental nongovernmental organizations, 26
environmental protection regulations, 57, 74
Environmental Working Group, 7
ethics: of Ammostro, 169; of the social collectivity, 116; transcorporeal, 71
European Agency for Safety and Health at Work, 179n8
European Commission, 2
European Union (EU): courts of the, 82; funds from the, 139; legislation of the, 2, 30; report of the, 188n49. *See also* Seveso Directive
European Union Court of Human Rights, 73

Favel, Greco, 56
feminism, 29; in film culture, 132; material, 71–72. *See also* feminist ecocriticism; women
feminist ecocriticism, 44. *See also* feminism
feminist new materialism, 11
Ferraris, Maurizio, 79–80
film: counterhegemonic, 113; documentary, 18, 108–12, 116, 125–26, 131–32, 136, 141, 189n5; environmentalist, 119, 132–34; feminist, 113, 132; horror, 189n4; international treatments of Seveso in, 60–61, 107; Italian, 109, 190n14; literature and, 80; observational, 112; political, 56; sensorial capacities of, 59. *See also* ecocinema; ecocritical film studies; ecodocumentary; media
Fisher-Lichte, Erika, 167
Flaherty, Robert: *Nanook of the North* (film), 109
Fondazione Corriere della Sera, 151
Fondazione Lombardia per l'Ambiente, 150
Fondazione Micheletti archives, 24, 40, 56
Fondo Antidiossina (Anti-Dioxin Fund), 69
Fornaro, Angelo, 67, 97, 127, 137, 162–64
Fornaro, Vincenzo, 130, 137–40, 162–64, 167–68
Foschini, Giuliano, 70, 100, 188n51; *Quindici passi*, 96–99, 122
Fosso del Ronchetto, 32
France, 146
Frank, Arthur, 13
Fratter, Massimiliano, 61–62, 152–53; *Seveso: Memorie da sotto il bosco*, 23–24, 148, 150, 193n16
Fuller, John G.: *The Poison That Fell from the Sky*, 23
furans, 7. *See also* chlorinated dibenzofurans (CDFs)

Galbiati, Clemente, 153
Garrone, Matteo: *Gomorra* (film), 80
Gazzano, Noel, 17, 161–62, 164–68; *L'Insopportabile contraddizione* (project), 165; *Terra Mia, Io Cammino per Te* (video record), 165
Genette, Gérard, 54
Genoa, 66
Germany, 8, 107
Giorno, Il, 24
Gisinger, Sabine: *Gambit* (documentary film), 60
Givaudan, 2, 21, 25, 146. *See also* La Roche

global warming, 102–3
Gramsci, Antonio, 118
Great Acceleration, 4
Greek mythology, 64
Guattari, Félix, 55
Guido, Fido, 127

Hagenbach, Otto, 105
Hall, Troy E., 158
Ham, Sam H., 158
Hayward, Eva, 71
health: corporeal, 2, 8, 23; disasters of, 17, 28, 30; environmental, 2, 69, 74; realities of, 5, 29. *See also* disease; health crisis
health crisis, 1, 30. *See also* eco-corporeal crisis; health
Health Office of the Greater Lombardy region (Regione Lombardia Assessorato alla sanità, Giunta regionale), 24
heavy metal particles, 69
Heise, Ursula, 15
hemp, 17, 161–62, 164, 166, 169–70; fabrics crafted from, 169, 176; as the opposite to dioxin, 167; planting of, 168
Herman, David, 14, 16, 45, 142
Hernàndez Rodrìguez, Àlvaro Ivàn, 168
history: Italian environmental, 16, 154; objective recounting of, 62; political, 156
Ho Chi Minh, 47
Hoffman La Roche. *See* La Roche
hope, 83, 100, 138
Houppert, Karen, 8
human microbiome, 11–12
Human Microbiome Project, 11
Hunter, Ross, 105–6, 189n2
Hutton, Peter, 120

ideology: of neorealism, 80; and power, 29; science and, 30
Ilva/Italsider, 2–6, 18, 65–98, 108–18, 126–44, 163, 167–73, 177, 184n1, 188n49; blast furnaces of, 136, 173; closure of, 67, 69, 171, 173; dioxin-rich emissions of, 70, 84–86; executives of, 67, 137, 179n1; smokestacks of, 101, 127, 129; trials of, 162, 174; workers of, 69, 90–92. *See also* industrial emissions; steelworks
Ilva Football Club (documentary film), 95
indigenous peoples, 135
industrial accident, 103; large-scale, 146. *See also* industry
industrial emissions: European environmental standards for, 2; industrial pollution and, 63; peak in reported, 83; toxic, 4, 86; widespread runaway, 71. *See also* Ilva/Italsider; industry; steelworks
industrialization, 101–2, 104. *See also* industry
Industrial Reconstruction Institute (IRI), 65
Industrie Chimiche Meda Società Azionaria (ICMESA), 2, 6–7, 25, 38–40, 57, 146–49; explosion at the, 47–49, 70, 78, 106, 145–46, 153, 159–60; officials of, 39, 60; site of, 4–5, 19, 21–25, 30–34, 47, 106, 142–43, 154–56
industry: agricultural, 4; extractive, 172; fishing, 4; Italian, 65, 102–4, 114, 118; production cycle of, 103; speed of, 168; and toxic exposure, 143. *See also* industrial accident; industrial emissions; industrialization; steelworks
infrastructure, 65
Ingram, David, 111
International Agency for Research on Cancer, 9
interpretation (*hermeneia*), 80
In viaggio con Cecilia (documentary film), 108, 112–14, 116–20, 123, 132
Iovino, Serenella, 14, 41, 44–45, 51–54, 143, 183n21; *Ecocriticism and Italy*, 16–17, 160
Ireland, 8
ISLE: Interdisciplinary Studies in Literature and Environment, 16, 44
Italian Communist Party, 37
Italian Council of Ministers, 66
Italianist, The, 16

Italsider steelworks. *See* Ilva/Italsider
Italy, 9, 75, 107, 144, 154; abortion debate in, 29; Campania region of, 8; economic boom in, 110, 113; government of, 96; industrial promise of, 65, 102, 104, 114, 118; map of, *3*; north-central, 141; northern, 21, 37, 45; Po River valley of, 141; postmillennial life in, 80; realist portrayals of the society of, 81, 113; southern, 21, 78, 118, 135; zombie and postapocalyptic genres of, 105. *See also* Seveso; Taranto
Ivakhiv, Adrian, 129

Jackson, J. B., 144, 155, 192n7
James, Erin, 14–15, 45, 54, 60, 111, 143, 190n21
Japan, 104, 107; dioxin emissions in, 8
Jurecic, Ann, 13

Karin B toxic waste incident, 57
Knickerbocker, Scott, 15, 111

labbro leporino (cleft lip), 48
Lackey, Brenda K., 158
landscape: belonging to the, 166; boundedness of, 144; coevolutionary, 41; contemporary Italian, 161; nonhuman, 26, 170–72, 192n7; polluted, 4; reading, 141–73; as text, 16, 19, 143; as theater, 167, 172; troubled, 167. *See also* embodiment; narrative
Landscapes, Natures, Ecologies: Italy and the Environmental Humanities (Iovino, Past, and Cesaretti), 16
La Roche, 2, 6, 21, 61, 146. *See also* Givaudan
Latour, Bruno, 142
Lecce, 120
Lefebvre, Martin, 129
Legambiente Lombardia, 150–53
Legamjonici, 83
Lehtimäki, Markku, 45, 54, 111
Lenzi, Umberto: *Nightmare City*, 17, 61, 105–7, 189n2
Leonardi, Emanuele, 68

Levinson, Barry: *Good Morning, Vietnam* (film), 121–22
Liboiron, Max, 6
Lombardy, 37, 193n13
love, 64
Love Canal crisis, 7–8. *See also* ecocorporeal crisis

MacDonald, Scott, 120; "Toward an Ecocinema," 110–11
macronarrative, 55
Made in ILVA (theater production), 77
Mamme da Nord a Sud (Moms from the North to the South), 166
Mangini, Cecilia, 109, 112–21; *La Briglia del collo*, 113; *Brindisi '65*, 113–14; *Due scatole dimenticate*, 112; *Essere donne*, 118–19; *Ignoti alla città*, 113; *Stendalì—Suonano ancora*, 113; *Tommaso*, 113–14
mapping, 79, 84–95; spatial, 94–95
Marescotti, Alessandro, 70–71
Mar Grande, 104, 129, 133
Mar Piccolo, 65, 102, 104, 144; living creatures of the, 169, 175
Marsella, Roberto, 77, 126, 130
Mascherpa, Barbara, 29
Masseria Carmine, 144, 162–65, 194n46
material ecocriticism, 41–42
materiality, 60
Meda, 22, 24, 147–48, 154, 156
media: international, 67, 69; mainstream, 5, 7; national, 31, 69; representational, 81; sensationalist coverage in the, 3. *See also* film; radio; television
MEDIMEX International Festival and Music Conference, 173
memory, 156; collective, 150–52; historical, 17, 19; loss of, 102, 153
Memory Bridge (Il Ponte della Memoria), 150–52, 160. *See also* Oak Forest (Seveso)
Memory Path (Percorso della Memoria), 151–58, 160. *See also* Oak Forest (Seveso)
micronarrative, 48–53, 55, 58–60; indirect, 78

Milan, 2, 25, 37, 40, 147, 154; Mangiagalli hospital in, 28; post-Seveso abortions in, 29; regional public health headquarters in, 25–26
Milan Provincial Laboratory for Hygiene and Prevention, 24
Missero, Dalila, 113, 118
Modena, 57
Montecatini-Shell petrochemical plant, 114, 118
Moran-Thomas, Amy, 27
Morel, Eric, 143
Mortimer-Sandilands, Catriona, 160

Naples, 93, 174
narrative: autobiographical, 17; contemporary realist, 85; and counter-narrative, 5; creative, 78, 143; diverse plural, 100; experience of, 16; liberatory capacity of, 17; multidimensional, 81; multisensorial, 81; ontological, 97; shaping of, 55, 143; Taranto-based, 78–79, 81, 89; traditional arc of, 16; verbal, 112; visual, 112. *See also* landscape; storyworld
narratology, 14, 109; cognitive, 45; contextualist, 45
Nash, Kate, 189n5
National Archaeological Museum of Taranto (MarTA), 64
National Institute for Health, 11
neoliberalism, 172
neorealism, 109
Netflix, 132
New Realism, 79–80
New York Times Magazine, 11
Nichols, Bill, 107–8, 189n5
Nigeria, 57
Nixon, Rob: *Slow Violence*, 3, 5
Non c'era nessuna signora a quel tavolo (biographical documentary film), 113
Non-Governmental Organization (NGO), 32
Non perdono (documentary film), 17, 77, 95, 108, 112, 126–28, 132, 164, 167
North Atlantic Treaty Organization (NATO), 131

novel: coming-of-age, 89; realist, 77
Nuova scuola letture, 56
Nussbaum, Martha, 50

Oak Forest (Seveso), 3, 17, 19, 62, 142–60; history of the, 160; landscape of the, 192n7. *See also* Memory Bridge (Il Ponte della Memoria); Memory Path (Percorso della Memoria)
O'Conner, Kevin, 168
O'Leary, Alan, 109
olives, 138
Ombre sulla città perduta (novel), 79
Opperman, Serpil, 41
Oracle of Delphi, 63
organized crime, 80

Pagano, Tullio, 144
Palumbo Mosca, Raffaele, 79–80
Paoletti, Paolo, 23
Paolucci, Giovanni: *Acciaio fra gli ulivi* (film), 121
Parteni, 63
Pasolini, Pier Paolo, 66
Past, Elena, 59–60, 114, 133–34, 145, 149, 171; *Italian Ecocinema beyond the Human*, 16–17, 186n18
Payne, Donald M., 7
PeaceLink, 69–70
pensiero debole (weak thought), 79
performance: melodramatic, 58; scholars of, 143; scripted, 141
Perniola, Ivelise, 122
Phelan, James, 109
Phoenician tradition, 64
phytoremediation, 162
Pianzola, Nicola, 77
Piccinni, Flavia, 127; *Adesso tienimi* (novel), 81–82, 84–90, 93, 97–99
Pignatelli, Luigi, 126–31
Pisa, 83–84
Pisanelli, Paolo: *Buongiorno Taranto*, 95, 101, 108, 112, 120–26, 132; *Due scatole dimenticate*, 112
plurivocality, 77–100, 124
politics, 113; local, 139

Pollan, Michael: "Some of My Best Friends Are Germs," 11
pollution: campaigns against, 32, 165; industrial, 18, 34, 63, 66–69, 74, 171; organic, 57–58; remediation of, 30; of Taranto, 86, 139; and toxicity, 99–100. *See also* toxicity
polychlorinated biphenyls (PCBs), 7–8, 57, 69; imprint of, 76. *See also* dioxin
Ponte della Memoria history project, 32
pope, 29. *See also* Catholic Church
Po River, 141
Porta Napoli, 103
postmodernism, 79
Povinelli, Elizabeth, 172
Powell, Kimberly, 144–45
protest: and community action, 69; public, 108; and survival, 63; by workers, 67–69
Prudenzano, Angelo, 89
public opinion, 4
Puglia, 78, 112, 114, 117, 119–20, 165–66; illness in, 166; institutional support for large-scale cultural events by, 173; tradition of pizzica (tarantella) music and dance of, 120

racism, 135
radio, 123–25. *See also* media
RAI (Italian national public broadcasting company), 109–10
Rame, Franca, 56–58
Re, Anna, 44; *Italian Environmental Literature: An Anthology*, 44
reading, 15
representation: Italian practices of, 80; of media, 81
Repubblica, La, 80, 82, 96, 98
Ricci, Barbara, 56
Ricciardi, Stefania, 79
Riondino, Michelle, 191n50
Riva, Emilio, 108, 117
Riva Group, 66–67, 96
Rivolta, Vittorio, 25
Rocca, Francesco, 24; *I giorni della diossina*, 23

Rome, 30
Routledge Encyclopedia of Narrative Theory, 15
Rubini, Sergio: *Il Grande Spirito*, 135–36

Sadeghian, Mohammad Mehdi, 155
Safran, Linda, 128
Salinella, 177. *See also* Taranto
Sambeth, Jorg, 60
Sandrelli, Amanda, 56
Sartori, Maria Luisa, 26–27
Sassi, Mauro, 108–9
Saviano, Roberto: *Gomorra* (novel), 80
Schoonover, Karl, 58, 60
Sciaje, Le, 101–2
science, 7, 35; environmental, 78; and ideology, 30
Seger, Monica: *Landscapes in Between: Environmental Change in Modern Italian Literature and Film*, 16
SENTIERI project, 68
Serra, Gianni, 18, 184n47, 184n50; *Una lepre con la faccia di bambina* (film), 56–60
Seveso, 2–13, 3, 15–50, 70, 106, 139, 142, 146–54; artisanal community of, 40; contaminated areas of, 25, 30; dioxins released in, 51, 145; landscape of, 144; Oak Forest of, 3, 17, 19, 62, 142–60; residents of, 56, 78, 183n38, 184n47; stories of, 37–61, 78. *See also* Italy; Oak Forest (Seveso)
Seveso Directive, 30. *See also* European Union (EU)
Seveso memorial park, 4–5
sexuality, 51; female, 48
Sheehi, Stephen, 187n40
Simonetti, Gianluigi, 85
Sinatra, Nancy: "These Boots Are Made for Walkin'" (song), 121
Sironi, Alberto, 141
slowness, 171–72
slow violence, 3, 5, 70. *See also* eco-corporeal crisis
Smith, Greg, 111, 119
Sobchack, Vivian, 59–60

socioenvironmental phenomena, 43
Soderbergh, Steven, 82
Sognando nuvole bianche (documentary film), 95, 98
soil, 22, 167–68; dioxin in samples of, 25, 180n5; pollution of, 166
Soldati, Mario: *Viaggio nella Valle del Po* (documentary film), 110
Soles, Carter, 106, 189n4
Somigli, Luca, 79
Sommer, Doris, 171
Spain, 107
Spera, Daniela, 70–71, 73–76, 82–83, 87, 101, 126, 137, 163, 174
Spirn, Anne Whiston, 143–44
Spiro, Thomas G., 22
steelworks, 2–4, 68–69, 88–98, 104, 117, 131, 168–70; dioxins from the, 75, 90, 125; physical space of the, 139; polluted soil closest to the Ilva, 166; toxins from the, 78, 90, 125; workers of the, 69, 90–92, 138. See also Ilva/Italsider; industrial emissions; industry
Steiglitz, Hugo, 106
Steingraber, Sandra, 13
Stockholm Convention on Persistent Organic Pollutants, 9
storyworld, 81, 100; environmentally situated, 81. See also narrative
Studies in Documentary Film (special issue), 109
subjectivity: human, 72; transversal, 71–72
sulfuric acid, 69
Swistun, Debora, 52
Switzerland, 25, 146

Tamburi, 3, 68–69, 75, 86, 91–93, 98, 103–4, 127, 169, 176; children of, 171. See also Taranto
Taranto, 2–19, 3, 61–104, 110–21, 128–35, 138–42, 160–73, 174–77; beauty of, 121, 125, 139, 171; city council of, 137; GDP of, 68; geography of, 79, 85, 130; graffiti culture of, 185n1; history of, 89, 99, 114, 185n4; illness in, 90, 99, 138, 173; industrial cluster of, 165; landscape of, 144, 165; maritime history of, 101; religious pageantry traditions of, 112; religious traditions of, 128–29; residents of, 71, 85–86, 94, 139, 171, 177; rooftops of, 135–36; toxic reality of, 100, 145; traditional forms of livelihood of, 170. See also Città Vecchia; Italy; Salinella; Tamburi
television, 18, 34, 58–59, 61, 183n47. See also media
Tempa Rossa oil fields, 65
Tempesta Films, 134
temporality: ambiguities of the toxic substance that pertain to, 29; differences rooted in, 4; interdependencies of, 4, 42; of the more-than-human world, 72; spatio-temporal relations and, 45–46, 142
theater, 167
Thomas, Valerie M., 22
Times Beach, 8
tourism, 139, 188n49
toxic chemical compounds, 4, 9–10, 146, 170; exposure to, 8, 68, 73–74, 143; imperceptibility of, 81; legal limits for, 73–74. See also dioxin; furans; polychlorinated biphenyls (PCBs); toxicity
toxic colonialism, 57
toxicity, 1, 4, 70, 85–90, 100; chemical, 46, 51, 183n38; in daily emissions, 67–68; dioxin, 7, 39, 45, 49–51, 54, 67, 90; ecocorporeal, 6; environmental, 17, 43–44, 94, 97; experience of, 88; and female sexuality, 48; Italian narratives of, 161; of liver, lungs, and kidneys, 28; nonvisible, 58; pollution and, 99; reports of, 47, 96; visibility of, 58. See also pollution; toxic chemical compounds; toxic waste
toxic waste, 57. See also toxicity
transcorporeality, 71, 89, 100, 166; clear explanation of, 166; dioxin-rich, 72; experiences of transmutability and, 167. See also embodiment
trichlorophenol, 21–22, 25
Trotter, Laura, 106
Tuan, Yi-Fu, 142–43

Tuana, Nancy, 87
turbocapitalist economy, 172. *See also* capitalism
Turri, Eugenio, 143, 159–60, 167
2,3,7,8-tetrachlorodibenzo-p-dioxin (2,3,7,8-TCDD), 7–8, 21–22, 29, 47–49, 156

Udine, 37
Unità, L' (newspaper), 145, 184n50
United Nations Environment Program (UNEP), 9
United States, 9; antitampon movements in the, 8
U.S. Department of Health and Human Services, 9
U.S. Environmental Protection Agency, 9

Vardanyan, Zhirayr, 155
Varedo, 34
Vattimo, Gianni, 79–80
Veneto, 21
Verdicchio, Pasquale: *Ecocritical Approaches to Italian Culture and Literature: The Denatured Wild*, 16

Vietnam War, 7, 21, 47–48
Villa Dho, 32
Village Voice, 8

walking, 168
Washington Times, 57
Williams, Robin, 121–22
Willoquet-Maricondi, Paula, 111
women: as artists, 168–69; breast milk of Tarantine, 7; caretaking responsibilities held by, 167; gender roles for southern Italian, 166; Seveso case as an issue for, 29. *See also* feminism
World Health Organization, 9, 180n5
World War I, 104
World War II, 21, 37, 65, 104, 163
World Wildlife Federation (WWF), 127

Yuschenko, Viktor, 9

Zagaria, Cristina: *Veleno*, 17, 73, 77, 82–84, 87–89, 93, 96
Zanotto, Grace, 77, 126, 130, 132

Recent books in the series
UNDER THE SIGN OF NATURE: EXPLORATIONS IN ECOCRITICISM

Taylor A. Eggan • *Unsettling Nature: Ecology, Phenomenology, and the Settler Colonial Imagination*

Samuel Amago • *Basura: Cultures of Waste in Contemporary Spain*

Marco Caracciolo • *Narrating the Mesh: Form and Story in the Anthropocene*

Tom Nurmi • *Magnificent Decay: Melville and Ecology*

Elizabeth Callaway • *Eden's Endemics: Narratives of Biodiversity on Earth and Beyond*

Alicia Carroll • *New Woman Ecologies: From Arts and Crafts to the Great War and Beyond*

Emily McGiffin • *Of Land, Bones, and Money: Toward a South African Ecopoetics*

Elizabeth Hope Chang • *Novel Cultivations: Plants in British Literature of the Global Nineteenth Century*

Christopher Abram • *Evergreen Ash: Ecology and Catastrophe in Old Norse Myth and Literature*

Serenella Iovino, Enrico Cesaretti, and Elena Past, editors • *Italy and the Environmental Humanities: Landscapes, Natures, Ecologies*

Julia E. Daniel • *Building Natures: Modern American Poetry, Landscape Architecture, and City Planning*

Lynn Keller • *Recomposing Ecopoetics: North American Poetry of the Self-Conscious Anthropocene*

Michael P. Branch and Clinton Mohs, editors • *"The Best Read Naturalist": Nature Writings of Ralph Waldo Emerson*

Jesse Oak Taylor • *The Sky of Our Manufacture: The London Fog in British Fiction from Dickens to Woolf*

Eric Gidal • *Ossianic Unconformities: Bardic Poetry in the Industrial Age*

Adam Trexler • *Anthropocene Fictions: The Novel in a Time of Climate Change*

Kate Rigby • *Dancing with Disaster: Environmental Histories, Narratives, and Ethics for Perilous Times*

Byron Caminero-Santangelo • *Different Shades of Green: African Literature, Environmental Justice, and Political Ecology*

Jennifer K. Ladino • *Reclaiming Nostalgia: Longing for Nature in American Literature*

Dan Brayton • *Shakespeare's Ocean: An Ecocritical Exploration*

Scott Hess • *William Wordsworth and the Ecology of Authorship: The Roots of Environmentalism in Nineteenth-Century Culture*

Axel Goodbody and Kate Rigby, editors • *Ecocritical Theory: New European Approaches*

Deborah Bird Rose • *Wild Dog Dreaming: Love and Extinction*

Paula Willoquet-Maricondi, editor • *Framing the World: Explorations in Ecocriticism and Film*

Bonnie Roos and Alex Hunt, editors • *Postcolonial Green: Environmental Politics and World Narratives*

Rinda West • *Out of the Shadow: Ecopsychology, Story, and Encounters with the Land*

Mary Ellen Bellanca • *Daybooks of Discovery: Nature Diaries in Britain, 1770–1870*

John Elder • *Pilgrimage to Vallombrosa: From Vermont to Italy in the Footsteps of George Perkins Marsh*

Alan Williamson • *Westernness: A Meditation*

Kate Rigby • *Topographies of the Sacred: The Poetics of Place in European Romanticism*

Mark Allister, editor • *Eco-Man: New Perspectives on Masculinity and Nature*

Heike Schaefer • *Mary Austin's Regionalism: Reflections on Gender, Genre, and Geography*

Scott Herring • *Lines on the Land: Writers, Art, and the National Parks*

Glen A. Love • *Practical Ecocriticism: Literature, Biology, and the Environment*

Ian Marshall • *Peak Experiences: Walking Meditations on Literature, Nature, and Need*

Robert Bernard Hass • *Going by Contraries: Robert Frost's Conflict with Science*

Michael A. Bryson • *Visions of the Land: Science, Literature, and the American Environment from the Era of Exploration to the Age of Ecology*

Ralph H. Lutts • *The Nature Fakers: Wildlife, Science, and Sentiment*